Y0-FZX-119

WITHDRAWN

INTRODUCTORY PROBABILITY THEORY

SELECTED RUSSIAN PUBLICATIONS IN THE MATHEMATICAL SCIENCES

Richard A. Silverman, *Editor*

ABRIKOSOV ET AL. *Methods of Quantum Field Theory in Statistical Physics*
BORISENKO & TARAPOV *Vector and Tensor Analysis with Applications*
GELFAND & FOMIN *Calculus of Variations*
LEBEDEV, *Special Functions and Their Applications*
LEBEDEV, SKALSKAYA & UFLYAND *Problems of Mathematical Physics*
MARKUSHEVICH *Theory of Functions of a Complex Variable, Volume I*
MARKUSHEVICH *Theory of Functions of a Complex Variable, Volume II*
MARKUSHEVICH *Theory of Functions of a Complex Variable, Volume III*
PETROVSKI *Ordinary Differential Equations*
ROZANOV *Introductory Probability Theory*
SHILOV *An Introduction to the Theory of Linear Spaces*
SHILOV & GUREVICH *Integral, Measure and Derivative: A Unified Approach*
TOLSTOV *Fourier Series*
WAINSTEIN & ZUBAKOV *Extraction of Signals from Noise*
YAGLOM *An Introduction to the Theory of Stationary Random Functions*

PRENTICE-HALL INTERNATIONAL, INC., *London*
PRENTICE-HALL OF AUSTRALIA, PTY. LTD., *Sydney*
PRENTICE-HALL OF CANADA, LTD., *Toronto*
PRENTICE-HALL OF INDIA PRIVATE LTD., *New Delhi*
PRENTICE-HALL OF JAPAN, INC., *Tokyo*

INTRODUCTORY PROBABILITY THEORY

Y. A. ROZANOV

Revised English Edition
Translated and Edited by

Richard A. Silverman

PRENTICE-HALL, INC.
Englewood Cliffs, New Jersey

241193

© 1969 by
PRENTICE-HALL, INC.
Englewood Cliffs, N.J.

All rights reserved. No part of this book may be reproduced in any form, or by any means without permission in writing from the publisher.

QA
273
R87413

Current Printing:

10 9 8 7 6 5 4 3 2 1

13-501932-X

Library of Congress Catalog Number 72-85666

Printed in the United States of America

EDITOR'S PREFACE

This book is a concise introduction to modern probability theory and certain of its ramifications. By deliberate succinctness of style and judicious selection of topics, it manages to be both fast-moving and self-contained.

The present edition differs from the Russian original (Moscow, 1968) in several respects:

1. It has been heavily restyled with the addition of some new material. Here I have drawn from my own background in probability theory, information theory, etc.

2. Each of the eight chapters and four appendices has been equipped with relevant problems, many accompanied by hints and answers. There are 150 of these problems, in large measure drawn from the excellent collection edited by A. A. Sveshnikov (Moscow, 1965).

3. At the end of the book I have added a brief Bibliography, containing suggestions for collateral and supplementary reading.

R. A. S.

CONTENTS

INTRODUCTORY PROBABILITY THEORY

1

BASIC CONCEPTS

1. Probability and Relative Frequency

Consider the simple experiment of tossing an unbiased coin. This experiment has two mutually exclusive outcomes, namely "heads" and "tails." The various factors influencing the outcome of the experiment are too numerous to take into account, at least if the coin tossing is "fair." Therefore the outcome of the experiment is said to be "random." Everyone would certainly agree that the "probability of getting heads" and the "probability of getting tails" both equal $\frac{1}{2}$. Intuitively, this answer is based on the idea that the two outcomes are "equally likely" or "equiprobable," because of the very nature of the experiment. But hardly anyone will bother at this point to clarify just what he means by "probability."

Continuing in this vein and taking these ideas at face value, consider an experiment with a finite number of mutually exclusive outcomes which are *equiprobable*, i.e., "equally likely because of the nature of the experiment." Let A denote some event associated with the possible outcomes of the experiment. Then the *probability* $\mathbf{P}(A)$ of the event A is defined as the fraction of the outcomes in which A occurs. More exactly,

$$\mathbf{P}(A) = \frac{N(A)}{N}, \tag{1.1}$$

where N is the total number of outcomes of the experiment and $N(A)$ is the number of outcomes leading to the occurrence of the event A.

***Example* 1.** In tossing a well-balanced coin, there are $N = 2$ mutually exclusive equiprobable outcomes ("heads" and "tails"). Let A be either of these two outcomes. Then $N(A) = 1$, and hence

$$\mathbf{P}(A) = \frac{1}{2}.$$

***Example* 2.** In throwing a single unbiased die, there are $N = 6$ mutually exclusive equiprobable outcomes, namely getting a number of spots equal to each of the numbers 1 through 6. Let A be the event consisting of getting an even number of spots. Then there are $N(A) = 3$ outcomes leading to the occurrence of A (which ones?), and hence

$$\mathbf{P}(A) = \frac{3}{6} = \frac{1}{2}.$$

***Example* 3.** In throwing a pair of dice, there are $N = 36$ mutually exclusive equiprobable events, each represented by an ordered pair (a, b), where a is the number of spots showing on the first die and b the number showing on the second die. Let A be the event that both dice show the same number of spots. Then A occurs whenever $a = b$, i.e., $n(A) = 6$. Therefore

$$\mathbf{P}(A) = \frac{6}{36} = \frac{1}{6}.$$

Remark. Despite its seeming simplicity, formula (1.1) can lead to nontrivial calculations. In fact, before using (1.1) in a given problem, we must find all the equiprobable outcomes, and then identify all those leading to the occurrence of the event A in question.

The accumulated experience of innumerable observations reveals a remarkable regularity of behavior, allowing us to assign a precise meaning to the concept of probability not only in the case of experiments with equiprobable outcomes, but also in the most general case. Suppose the experiment under consideration can be repeated any number of times, so that, in principle at least, we can produce a whole series of "independent trials under identical conditions,"[1] in each of which, depending on chance, a particular event A of interest either occurs or does not occur. Let n be the total number of experiments in the whole series of trials, and let $n(A)$ be the number of experiments in which A occurs. Then the ratio

$$\frac{n(A)}{n}$$

is called the *relative frequency* of the event A (in the given series of trials). It turns out that the relative frequencies $n(A)/n$ observed in different series of

---

[1] Concerning the notion of independence, see Sec. 6, in particular footnote 2, p. 31.

trials are virtually the same for large n, clustering about some constant

$$\mathbf{P}(A) \sim \frac{n(A)}{n}, \tag{1.2}$$

called the *probability* of the event A. More exactly, (1.2) means that

$$\mathbf{P}(A) = \lim_{n\to\infty} \frac{n(A)}{n}. \tag{1.3}$$

Roughly speaking, the probability $\mathbf{P}(A)$ of the event A equals the fraction of experiments leading to the occurrence of A in a large series of trials.[2]

***Example* 4.** Table 1 shows the results of a series of 10,000 coin tosses,[3] grouped into 100 different series of $n = 100$ tosses each. In every case, the table shows the number of tosses $n(A)$ leading to the occurrence of a head. It is clear that the relative frequency of occurrence of "heads" in each set of 100 tosses differs only slightly from the probability $\mathbf{P}(A) = \frac{1}{2}$ found in Example 1. Note that the relative frequency of occurrence of "heads" is even closer to $\frac{1}{2}$ if we group the tosses in series of 1000 tosses each.

Table 1. Number of heads in a series of coin tosses

Number of heads in 100 series of 100 trials each										*Number of heads in 10 series of 1000 trials each*[4]
54	46	53	55	46	54	41	48	51	53	501
48	46	40	53	49	49	48	54	53	45	485
43	52	58	51	51	50	52	50	53	49	509
58	60	54	55	50	48	47	57	52	55	536
48	51	51	49	44	52	50	46	53	41	485
49	50	45	52	52	48	47	47	47	51	488
45	47	41	51	49	59	50	55	53	50	500
53	52	46	52	44	51	48	51	46	54	497
45	47	46	52	47	48	59	57	45	48	494
47	41	51	48	59	51	52	55	39	41	484

***Example* 5 (*De Méré's paradox*).** As a result of extensive observation of dice games, the French gambler de Méré noticed that the total number of spots showing on three dice thrown simultaneously turns out to be 11 (the event A_1) more often than it turns out to be 12 (the event A_2), although from his point of view both events should occur equally often. De Méré

[2] For a more rigorous discussion of the meaning of (1.2) and (1.3), see Sec. 12 on the "law of large numbers."

[3] Table 1 is taken from W. Feller, *An Introduction to Probability Theory and Its Applications, Volume I*, third edition, John Wiley and Sons, Inc., New York (1968), p. 21, and actually stems from a table of "random numbers."

[4] Obtained by adding the numbers on the left, row by row.

reasoned as follows: A_1 occurs in just six ways (6:4:1, 6:3:2, 5:5:1, 5:4:2, 5:3:3, 4:4:3), and A_2 also occurs in just six ways (6:5:1, 6:4:2, 6:3:3, 5:5:2, 5:4:3, 4:4:4). Therefore A_1 and A_2 have the same probability $\mathbf{P}(A_1) = \mathbf{P}(A_2)$.

The fallacy in this argument was found by Pascal, who showed that the outcomes listed by de Méré are not actually equiprobable. In fact, one must take account not only of the numbers of spots showing on the dice, but also of the particular dice on which the spots appear. For example, numbering the dice and writing the number of spots in the corresponding order, we find that there are six distinct outcomes leading to the combination 6:4:1, namely (6, 4, 1), (6, 1, 4), (4, 6, 1), (4, 1, 6), (1, 6, 4) and (1, 4, 6), whereas there is only one outcome leading to the combination 4:4:4, namely (4, 4, 4). The appropriate equiprobable outcomes are those described by triples of numbers (a, b, c), where a is the number of spots on the first die, b the number of spots on the second die, and c the number of spots on the third die. It is easy to see that there are then precisely $N = 6^3 = 216$ equiprobable outcomes. Of these, $N(A_1) = 27$ are favorable to the event A_1 (in which the sum of all the spots equals 11), but only $N(A_2) = 25$ are favorable to the event A_2 (in which the sum of all the spots equals 12).[5] This fact explains the tendency observed by de Méré for 11 spots to appear more often than 12.

2. Rudiments of Combinatorial Analysis

Combinatorial formulas are of great use in calculating probabilities. We now derive the most important of these formulas.

THEOREM 1.1. *Given n_1 elements $a_1, a_2, \ldots, a_{n_1}$ and n_2 elements $b_1, b_2, \ldots, b_{n_2}$, there are precisely $n_1 n_2$ distinct ordered pairs (a_i, b_j) containing one element of each kind.*

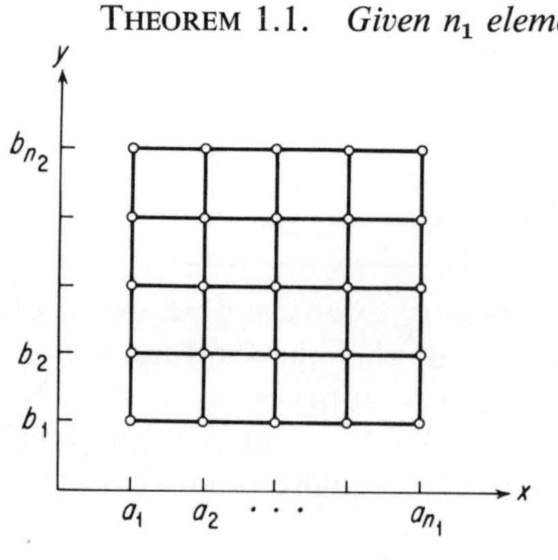

FIGURE 1.

Proof. Represent the elements of the first kind by points of the x-axis, and those of the second kind by points of the y-axis. Then the possible pairs (a_i, b_j) are points of a rectangular lattice in the xy-plane, as shown in Figure 1. The fact that there are just $n_1 n_2$ such pairs is obvious from the figure. ▌[6]

[5] To see this, note that a combination $a{:}b{:}c$ occurs in 6 distinct ways if a, b and c are distinct, in 3 distinct ways if two (and only two) of the numbers a, b and c are distinct, and in only 1 way if $a = b = c$. Hence A_1 occurs in $6 + 6 + 3 + 6 + 3 + 3 = 27$ ways, while A_2 occurs in $6 + 6 + 3 + 3 + 6 + 1 = 25$ ways.

[6] The symbol ▌ stands for Q.E.D. and indicates the end of a proof.

More generally, we have

THEOREM 1.2. *Given* n_1 *elements* $a_1, a_2, \ldots, a_{n_1}$, n_2 *elements* $b_1, b_2, \ldots, b_{n_2}$, *etc., up to* n_r *elements* $x_1, x_2, \ldots, x_{n_r}$, *there are precisely* $n_1 n_2 \cdots n_r$ *distinct ordered r-tuples* $(a_{i_1}, b_{i_2}, \ldots, x_{i_r})$ *containing one element of each kind.*[7]

Proof. For $r = 2$, the theorem reduces to Theorem 1.1. Suppose the theorem holds for $r - 1$, so that in particular there are precisely $n_2 \cdots n_r$ $(r - 1)$-tuples $(b_{i_2}, \ldots, x_{i_r})$ containing one element of each kind. Then, regarding the $(r - 1)$-tuples as elements of a new kind, we note that each r-tuple $(a_{i_1}, b_{i_2}, \ldots, x_{i_r})$ can be regarded as made up of a $(r - 1)$-tuple $(b_{i_2}, \ldots, x_{i_r})$ and an element a_{i_1}. Hence, by Theorem 1.1, there are precisely

$$n_1(n_2 \cdots n_r) = n_1 n_2 \cdots n_r$$

r-tuples containing one element of each kind. The theorem now follows for all r by mathematical induction. ∎

***Example* 1.** What is the probability of getting three sixes in a throw of three dice?

Solution. Let a be the number of spots on the first die, b the number of spots on the second die, and c the number of spots on the third die. Then the result of throwing the dice is described by an ordered triple (a, b, c), where each element takes values from 1 to 6. Hence, by Theorem 1.2 with $r = 3$ and $n_1 = n_2 = n_3 = 6$, there are precisely $N = 6^3 = 216$ equiprobable outcomes of throwing three dice (this fact was anticipated in Example 5, p. 3). Three sixes can occur in only one way, i.e., when $a = b = c = 6$. Therefore the probability of getting three sixes is $\frac{1}{216}$.

***Example* 2 (*Sampling with replacement*).** Suppose we choose r objects in succession from a "population" (i.e., set) of n distinct objects $a_1, a_2, \ldots, a_n$, in such a way that after choosing each object and recording the choice, we return the object to the population before making the next choice. This gives an "ordered sample" of the form

$$(a_{i_1}, a_{i_2}, \ldots, a_{i_r}). \tag{1.4}$$

Setting $n_1 = n_2 = \cdots = n_r = n$ in Theorem 1.2, we find that there are precisely

$$N = n^r \tag{1.5}$$

distinct ordered samples of the form (1.4).[8]

---

[7] Two ordered r-tuples $(a_{i1}, b_{i2}, \ldots, x_{ir})$ and $(a_{j1}, b_{j2}, \ldots, x_{jr})$ are said to be *distinct* if the elements of at least one pair a_{i1} and a_{j1}, b_{i2} and b_{j2}, $\ldots$, a_{ir} and b_{jr} are distinct.

[8] Two "ordered samples" $(a_{i1}, a_{i2}, \ldots, a_{ir})$ and $(a_{j1}, a_{j2}, \ldots, a_{jr})$ are said to be *distinct* if $a_{i_j} \neq a_{j_k}$ for at least one $k = 1, 2, \ldots, r$. This is a special case of the definition in footnote 7.

***Example* 3 (*Sampling without replacement*).** Next suppose we choose r objects in succession from a population of n distinct objects $a_1, a_2, \ldots, a_n$, in such a way that an object once chosen is removed from the population. Then we again get an ordered sample of the form (1.4), but now there are $n - 1$ objects left after the first choice, $n - 2$ objects left after the second choice, and so on. Clearly this corresponds to setting

$$n_1 = n, \quad n_2 = n - 1, \ldots, \quad n_r = n - r + 1$$

in Theorem 1.2. Hence, instead of n^r distinct samples as in the case of sampling with replacement, there are now only

$$N = n(n - 1) \cdots (n - r + 1) \tag{1.6}$$

distinct samples. If $r = n$, then (1.6) reduces to

$$N = n(n - 1) \cdots 2 \cdot 1 = n!\,, \tag{1.7}$$

the total number of *permutations* of n objects.

***Example* 4.** Suppose we place r distinguishable objects into n different "cells" ($r \leqslant n$), with no cell allowed to contain more than one object. Numbering both the objects and the cells, let i_1 be the number of the cell into which the first object is placed, i_2 the number of the cell into which the second object is placed, and so on. Then the arrangement of the objects in the cells is described by an ordered r-tuple $(i_1, i_2, \ldots, i_r)$. Clearly, there are $n_1 = n$ empty cells originally, $n_2 = n - 1$ empty cells after one cell has been occupied, $n_3 = n - 2$ empty cells after two cells have been occupied, and so on. Hence, the total number of distinct arrangements of the objects in the cells is again given by formula (1.6).

***Example* 5.** A subway train made up of n cars is boarded by r passengers ($r \leqslant n$), each entering a car completely at random. What is the probability of the passengers all ending up in different cars?

Solution. By hypothesis, every car has the same probability of being entered by a given passenger. Numbering both the passengers and the cars, let i_1 be the number of the car entered by the first passenger, i_2 the number of the car entered by the second passenger, and so on. Then the arrangement of the passengers in the cars is described by an ordered r-tuple $(i_1, i_2, \ldots, i_r)$, where each of the numbers $i_1, i_2, \ldots, i_r$ can range from 1 to n. This is equivalent to sampling with replacement, and hence, by Example 2, there are

$$N = n^r$$

distinct equiprobable arrangements of the passengers in the cars. Let A be the event that "no more than one passenger enters any car." Then A occurs if and only if all the numbers $i_1, i_2, \ldots, i_r$ are distinct. In other words, if A is to occur, the first passenger can enter one of n cars, but the second passenger can only enter one of $n - 1$ cars, the third passenger one of $n - 2$ cars, and so on. This is equivalent to sampling without replacement, and hence, by Example 3, there are

$$N(A) = n(n-1)\cdots(n-r+1)$$

arrangements of passengers in the cars leading to the occurrence of A. Therefore, by (1.1), the probability of A occurring, i.e., of the passengers all ending up in different cars, is just

$$\mathbf{P}(A) = \frac{n(n-1)\cdots(n-r+1)}{n^r}.$$

Any set of r elements chosen from a population of n elements, *without regard for order*, is called a *subpopulation of size r* of the original population. The number of such subpopulations is given by

THEOREM 1.3. *A population of n elements has precisely*

$$C_r^n = \frac{n!}{r!\,(n-r)!} \tag{1.8}$$

subpopulations of size $r \leqslant n$.

Proof. If order mattered, then the elements of each subpopulation could be arranged in $r!$ distinct ways (recall Example 3). Hence there are $r!$ times more "ordered samples" of r elements than subpopulations of size r. But there are precisely $n(n-1)\cdots(n-r+1)$ such ordered samples (by Example 3 again), and hence just

$$\frac{n(n-1)\cdots(n-r+1)}{r!} = \frac{n!}{r!\,(n-r)!}$$

subpopulations of size r. ∎

Remark. An expression of the form (1.8) is called a *binomial coefficient*, often denoted by

$$\binom{n}{r}$$

instead of C_r^n. The number C_r^n is sometimes called the *number of combinations of n things taken r at a time* (without regard for order).

The natural generalization of Theorem 1.3 is given by

THEOREM 1.4. *Given a population of n elements, let* $n_1, n_2, \ldots, n_k$ *be positive integers such that*

$$n_1 + n_2 + \cdots + n_k = n.$$

Then there are precisely

$$N = \frac{n!}{n_1!\, n_2! \cdots n_k!} \tag{1.9}$$

ways of partitioning the population into k subpopulations, of sizes $n_1, n_2, \ldots, n_k$, respectively.

Proof. The order of the subpopulations matters in the sense that $n_1 = 2,\ n_2 = 4, n_3, \ldots, n_k$ and $n_1 = 4,\ n_2 = 2, n_3, \ldots, n_k$ (say) represent different partitions, but the order of elements within the subpopulations themselves is irrelevant. The partitioning can be effected in stages, as follows: First we form a group of n_1 elements from the original population. This can be done in

$$N_1 = C^n_{n_1}$$

ways. Then we form a group of n_2 elements from the remaining $n - n_1$ elements. This can be done in

$$N_2 = C^{n-n_1}_{n_2}$$

ways. Proceeding in this fashion, we are left with $n - n_1 - \cdots - n_{k-2} = n_{k-1} + n_k$ elements after $k - 2$ stages. These elements can be partitioned into two groups, one containing n_{k-1} elements and the other n_k elements, in

$$N_{k-1} = C^{n-n_1-\cdots-n_{k-2}}_{n_{k-1}}$$

ways. Hence, by Theorem 1.2, there are

$$\begin{aligned} N &= N_1 N_2 \cdots N_{k-1} \\ &= C^n_{n_1} C^{n-n_1}_{n_2} \cdots C^{n-n_1-\cdots-n_{k-2}}_{n_{k-1}} \end{aligned}$$

distinct ways of partitioning the given population into the indicated k subpopulations. But

$$\begin{aligned} & C^n_{n_1} C^{n-n_1}_{n_2} \cdots C^{n-n_1-\cdots-n_{k-1}}_{n_{k-1}} \\ &= \frac{n!}{n_1!\,(n-n_1)!} \frac{(n-n_1)!}{n_2!\,(n-n_1-n_2)!} \cdots \frac{(n-n_1-\cdots-n_{k-2})!}{n_{k-1}!\,(n-n_1-\cdots-n_{k-2}-n_{k-1})!} \\ &= \frac{n!}{n_1!\,(n-n_1)!} \frac{(n-n_1)!}{n_2!\,(n-n_1-n_2)!} \cdots \frac{(n-n_1-\cdots-n_{k-2})!}{n_{k-1}!\, n_k!} \\ &= \frac{n!}{n_1!\, n_2! \cdots n_k!}, \end{aligned}$$

in keeping with (1.9). ▮

Remark. Theorem 1.4 reduces to Theorem 1.3 if

$$k = 2, \quad n_1 = r, \quad n_2 = n - r.$$

The numbers (1.9) are called *multinomial coefficients*, and generalize the binomial coefficients (1.8).

***Example* 6 (*Quality control*).** A batch of 100 manufactured items is checked by an inspector, who examines 10 items selected at random. If none of the 10 items is defective, he accepts the whole batch. Otherwise, the batch is subjected to further inspection. What is the probability that a batch containing 10 defective items will be accepted?

Solution. The number of ways of selecting 10 items out of a batch of 100 items equals the number of combinations of 100 things taken 10 at a time, and is just

$$N = C_{10}^{100} = \frac{100!}{10!\,90!}.$$

By hypothesis, these combinations are all equiprobable (the items being selected "at random"). Let A be the event that "the batch of items is accepted by the inspector." Then A occurs whenever all 10 items belong to the set of 90 items of acceptable quality. Hence the number of combinations favorable to A is

$$N(A) = C_{10}^{90} = \frac{90!}{10!\,80!}.$$

It follows from (1.1) that the probability of the event A, i.e., of the batch being accepted, equals[9]

$$\mathbf{P}(A) = \frac{N(A)}{N} = \frac{90!\,90!}{80!\,100!} = \frac{81 \cdot 82 \cdots 90}{91 \cdot 92 \cdots 100} \approx \left(1 - \frac{1}{10}\right)^{10} \approx \frac{1}{e},$$

where $e = 2.718\ldots$ is the base of the natural logarithms.

***Example* 7.** What is the probability that two playing cards picked at random from a full deck are both aces?

Solution. A full deck consists of 52 cards, of which 4 are aces. There are

$$C_2^{52} = \frac{52!}{2!\,50!} = 1326$$

ways of selecting a pair of cards from the deck. Of these 1326 pairs, there are

$$C_2^4 = \frac{4!}{2!\,2!} = 6$$

[9] The symbol $\approx$ means "is approximately equal to."

consisting of two aces. Hence the probability of picking two aces is just

$$\frac{C_2^4}{C_2^{52}} = \frac{6}{1326} = \frac{1}{221}.$$

***Example* 8.** What is the probability that each of four bridge players holds an ace?

Solution. Applying Theorem 1.4 with $n = 52$ and $n_1 = n_2 = n_3 = n_4 = 13$, we find that there are

$$\frac{52!}{13!\,13!\,13!\,13!}$$

distinct deals of bridge. There are $4! = 24$ ways of giving an ace to each player, and then the remaining 48 cards can be dealt out in

$$\frac{48!}{12!\,12!\,12!\,12!}$$

distinct ways. Hence there are

$$24\,\frac{48!}{(12!)^4}$$

distinct deals of bridge such that each player receives an ace. Therefore the probability of each player receiving an ace is just

$$24\,\frac{48!}{(12!)^4}\,\frac{(13!)^4}{52!} = \frac{24(13)^4}{52 \cdot 51 \cdot 50 \cdot 49} \approx 0.105.$$

Remark. Most of the above formulas contain the quantity

$$n! = n(n-1) \cdots 2 \cdot 1,$$

called n *factorial.* For large n, it can be shown that[10]

$$n! \sim \sqrt{2\pi n}\, n^n e^{-n}.$$

This simple asymptotic representation of $n!$ is known as *Stirling's formula.*[11]

PROBLEMS

1. A four-volume work is placed in random order on a bookshelf. What is the probability of the volumes being in proper order from left to right or from right to left?

[10] The symbol $\sim$ between two variables α_n and β_n means that the ratio $\alpha_n/\beta_n \to 1$ as $n \to \infty$.

[11] Proved, for example, in D. V. Widder, *Advanced Calculus*, second edition, Prentice-Hall, Inc., Englewood Cliffs, N.J. (1961), p. 386.

2. A wooden cube with painted faces is sawed up into 1000 little cubes, all of the same size. The little cubes are then mixed up, and one is chosen at random What is the probability of its having just 2 painted faces?

Ans. 0.096.

3. A batch of n manufactured items contains k defective items. Suppose m items are selected at random from the batch. What is the probability that l of these items are defective?

4. Ten books are placed in random order on a bookshelf. Find the probability of three given books being side by side.

Ans. $\frac{1}{15}$.

5. One marksman has an 80% probability of hitting a target, while another has only a 70% probability of hitting the target. What is the probability of the target being hit (at least once) if both marksman fire at it simultaneously?

Ans. 0.94.

6. Suppose n people sit down at random and independently of each other in an auditorium containing $n + k$ seats. What is the probability that m seats specified in advance ($m \leqslant n$) will be occupied?

7. Three cards are drawn at random from a full deck. What is the probability of getting a three, a seven and an ace?

8. What is the probability of being able to form a triangle from three segments chosen at random from five line segments of lengths 1, 3, 5, 7 and 9?

Hint. A triangle cannot be formed if one segment is longer than the sum of the other two.

9. Suppose a number from 1 to 1000 is selected at random. What is the probability that the last two digits of its cube are both 1?

Hint There is no need to look through a table of cubes.

Ans. $\frac{1}{10}$.

10. Find the probability that a randomly selected positive integer will give a number ending in 1 if it is

a) Squared;

b) Raised to the fourth power;

c) Multiplied by an arbitrary positive integer.

Hint. It is enough to consider one-digit numbers.

Ans. a) 0.2; b) 0.4; c) 0.04.

11. One of the numbers 2, 4, 6, 7, 8, 11, 12 and 13 is chosen at random as the numerator of a fraction, and then one of the remaining numbers is chosen at random as the denominator of the fraction. What is the probability of the fraction being in lowest terms?

Ans. $\frac{9}{14}$.

12. The word "drawer" is spelled with six scrabble tiles. The tiles are then randomly rearranged. What is the probability of the rearranged tiles spelling the word "reward?"

Ans. $\frac{1}{360}$.

13. In throwing $6n$ dice, what is the probability of getting each face n times? Use Stirling's formula to estimate this probability for large n.

14. A full deck of cards is divided in half at random. Use Stirling's formula to estimate the probability that each half contains the same number of red and black cards.

Ans. $$\frac{C_{13}^{26}C_{13}^{26}}{C_{26}^{52}} \approx \frac{2}{\sqrt{26\pi}} \approx 0.22.$$

15. Use Stirling's formula to estimate the probability that all 50 states are represented in a committee of 50 senators chosen at random.

16. Suppose $2n$ customers stand in line at a box office, n with 5-dollar bills and n with 10-dollar bills. Suppose each ticket costs 5 dollars, and the box office has no money initially. What is the probability that none of the customers has to wait for change?[12]

Ans. $$\frac{C_n^{2n} - C_{n+1}^{2n}}{C_n^{2n}} = \frac{1}{n+1}.$$

17. Prove that

$$\sum_{k=0}^{n} (C_k^n)^2 = C_n^{2n}.$$

Hint. Use the binomial theorem to calculate the coefficient of x^n in the product $(1 + x)^n(1 + x)^n = (1 + x)^{2n}$.

---

[12] A detailed solution is given in B. V. Gnedenko, *The Theory of Probability*, fourth edition (translated by B. D. Seckler), Chelsea Publishing Co., New York (1967), p. 43.

2

COMBINATION OF EVENTS

3. Elementary Events. The Sample Space

The mutually exclusive outcomes of a random experiment (like throwing a pair of dice) will be called *elementary events* (or *sample points*), and a typical elementary event will be denoted by the Greek letter ω. The set of all elementary events ω associated with a given experiment will be called the *sample space* (or *space of elementary events*), denoted by the Greek letter Ω. An event A is said to be "associated with the elementary events of Ω" if, given any ω in Ω, we can always decide whether or not ω leads to the occurrence of A. The same symbol A will be used to denote both the event A and the set of elementary events leading to the occurrence of A. Clearly, an event A occurs if and only if one of the elementary events ω in the set A occurs. Thus, instead of talking about the occurrence of the original event A, we can just as well talk about the "occurrence of an elementary event ω in the set A." From now on, we will not distinguish between an event associated with a given experiment and the corresponding set of elementary events, it being understood that all our events are of the type described by saying "one of the elementary events in the set A occurs." With this interpretation, events are nothing more or less than subsets of some underlying sample space Ω. Thus the *certain* (or *sure*) event, which always occurs regardless of the outcome of the experiment, is formally identical with the whole space Ω, while the *impossible event* is just the empty set $\varnothing$, containing none of the elementary events ω.

Given two events A_1 and A_2, suppose A_1 occurs if and only if A_2 occurs. Then A_1 and A_2 are said to be *identical* (or *equivalent*), and we write $A_1 = A_2$.

Example 1. In throwing a pair of dice, let A_1 be the event that "the total number of spots is even" and A_2 the event that "both dice turn up even or both dice turn up odd."[1] Then $A_1 = A_2$.

Example 2. In throwing three dice, let A_1 again be the event that "the total number of spots is even" and A_2 the event that "all three dice have either an even number of spots or an odd number of spots." Then $A_1 \neq A_2$.

Two events A_1 and A_2 are said to be *mutually exclusive* or *incompatible* if the occurrence of one event precludes the occurrence of the other, i.e., if A_1 and A_2 cannot occur simultaneously.

By the *union* of two events A_1 and A_2, denoted by $A_1 \cup A_2$, we mean the event consisting of the occurrence of at least one of the events A_1 and A_2. The union of several events $A_1, A_2, \ldots$ is defined in the same way, and is denoted by $\bigcup_k A_k$.

By the *intersection* of two events A_1 and A_2, denoted by $A_1 \cap A_2$ or simply by A_1A_2, we mean the event consisting of the occurrence of both events A_1 and A_2. By the intersection of several events $A_1, A_2, \ldots$, denoted by $\bigcap_k A_k$, we mean the event consisting of the occurrence of all the events $A_1, A_2, \ldots$.

Given two events A_1 and A_2, by the difference $A_1 - A_2$ we mean the event in which A_1 occurs but not A_2. By the *complementary event* of an event A,[2] denoted by $\bar{A}$, we mean the event "A does not occur." Clearly,

$$\bar{A} = \Omega - A.$$

Example 3. In throwing a pair of dice, let A be the event that "the total number of spots is even," A_1 the event that "both dice turn up even," and A_2 the event that "both dice turn up odd." Then A_1 and A_2 are mutually exclusive, and clearly

$$A = A_1 \cup A_2, \quad A_1 = A - A_2, \quad A_2 = A - A_1.$$

Let $\bar{A}$, $\bar{A}_1$ and $\bar{A}_2$ be the events complementary to A, A_1 and A_2, respectively. Then $\bar{A}$ is the event that "the total number of spots is odd," $\bar{A}_1$ the event that "at least one die turns up odd," and $\bar{A}_2$ the event that "at least one die turns up even." It is easy to see that

$$\bar{A}_1 - \bar{A} = \bar{A}_1 \cap A = A_2, \qquad \bar{A}_2 - \bar{A} = \bar{A}_2 \cap A = A_1.$$

The meaning of concepts like the union of two events, the intersection of two events, etc., is particularly clear if we think of events as sets of elementary events ω, in the way described above. With this interpretation,

[1] To "turn up even" means to show an even number of spots, and similarly for to "turn up odd."

[2] Synonymously, the "complement of A" or the "event complementary to A."

given events A_1, A_2 and A, $A_1 \cup A_2$ is the union of the *sets* A_1 and A_2, $A_1 \cap A_2$ is the intersection of the *sets* A_1 and A_2, $\bar{A} = \Omega - A$ is the complement of the *set* A relative to the whole space Ω, and so on. Thus the symbols $\cup$, $\cap$, etc. have their customary set-theoretic meaning. Moreover, the statement that "the occurrence of the event A_1 implies that of the event A_2" (or simply, "A_1 implies A_2") means that $A_1 \subset A_2$, i.e., that the set A_1 is a subset of the set A_2.[3]

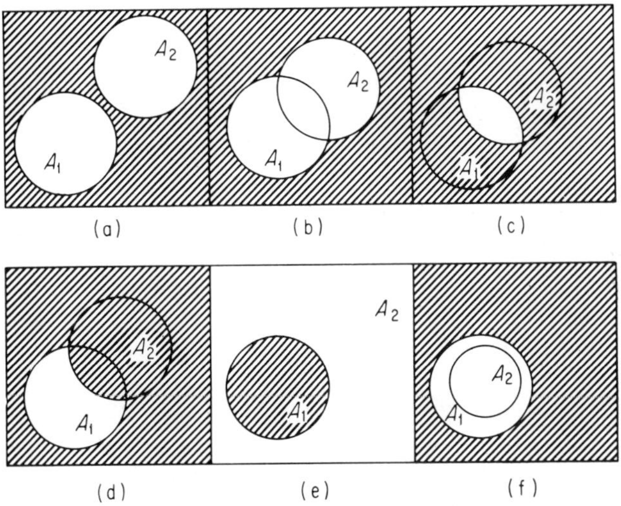

FIGURE 2. (a) The events A_1 and A_2 are mutually exclusive; (b) The unshaded figure represents the union $A_1 \cup A_2$; (c) The unshaded figure represents the intersection $A_1 \cap A_2$; (d) The unshaded figure represents the difference $A_1 - A_2$; (e) The shaded and unshaded events (A_1 and A_2) are complements of each other; (f) Event A_1 implies event A_2.

To visualize relations between events, it is convenient to represent the sample space Ω schematically by some plane region and the elementary events ω by points in this region. Then events, i.e., sets of points ω, become various plane figures. Thus Figure 2 shows various relations between two events A_1 and A_2, represented by circular disks lying inside a rectangle Ω, schematically representing the whole sample space. In turn, this way of representing events in terms of plane figures can be used to deduce general relations between events, e.g.,

a) If $A_1 \subset A_2$, then $\bar{A}_1 \supset \bar{A}_2$;
b) If $A = A_1 \cup A_2$, then $\bar{A} = \bar{A}_1 \cap \bar{A}_2$;
c) If $A = A_1 \cap A_2$, then $\bar{A} = \bar{A}_1 \cup \bar{A}_2$.

---

[3] The symbol $\subset$ means "is a subset of" or "is contained in," while $\supset$ means "contains."

Quite generally, given a relation between various events, we can get an equivalent relation by changing events to their complements and the symbols $\cap$, $\cup$ and $\subset$ to $\cap$, $\cup$ and $\supset$ (the sign $=$ is left alone).

***Example* 4.** The following relations are equivalent:

$$\bigcup_k A_k = B \subset \overline{\bigcap_k C_k},$$
$$\bigcap_k A_k = \bar{B} \supset \bigcap_k C_k,$$
$$\bigcup_k A_k = B \subset \bigcup_k \bar{C}_k.$$

Remark. It will henceforth be assumed that all events under consideration have well-defined probabilities. Moreover, it will be assumed that all events obtained from a given sequence of events $A_1, A_2, \ldots$ by taking unions, intersections, differences and complements also have well-defined probabilities.

4. The Addition Law for Probabilities

Consider two mutually exclusive events A_1 and A_2 associated with the outcomes of some random experiment, and let $A = A_1 \cup A_2$ be the union of the two events. Suppose we repeat the experiment a large number of times, thereby producing a whole series of "independent trials under identical conditions." Let n be the total number of trials, and let $n(A_1)$, $n(A_2)$ and $n(A)$ be the numbers of trials leading to the events A_1, A_2 and A, respectively. If A occurs in a trial, then either A_1 occurs or A_2 occurs, but not both (since A_1 and A_2 are mutually exclusive). Therefore

$$n(A) = n(A_1) + n(A_2),$$

and hence

$$\frac{n(A)}{n} = \frac{n(A_1)}{n} + \frac{n(A_2)}{n}.$$

But for sufficiently large n, the relative frequencies $n(A)/n$, $n(A_1)/n$ and $n(A_2)/n$ virtually coincide with the corresponding probabilities $\mathbf{P}(A)$, $\mathbf{P}(A_1)$ and $\mathbf{P}(A_2)$, as discussed on p. 3. It follows that

$$\mathbf{P}(A) = \mathbf{P}(A_1) + \mathbf{P}(A_2). \tag{2.1}$$

Similarly, if the events A_1, A_2 and A_3 are mutually exclusive, then so are $A_1 \cup A_2$ and A_3, and hence, by two applications of (2.1),

$$\mathbf{P}(A_1 \cup A_2 \cup A_3) = \mathbf{P}(A_1 \cup A_2) + \mathbf{P}(A_3) = \mathbf{P}(A_1) + \mathbf{P}(A_2) + \mathbf{P}(A_3).$$

More generally, given n mutually exclusive events $A_1, A_2, \ldots, A_n$, we have the formula

$$\mathbf{P}\left(\bigcup_{k=1}^{n} A_k\right) = \sum_{k=1}^{n} \mathbf{P}(A_k), \tag{2.2}$$

obtained by applying (2.1) $n - 1$ times. Equation (2.2) is called the *addition law for probabilities.*

Next we prove some key relations involving probabilities:

THEOREM 2.1. *The formulas*

$$0 \leqslant \mathbf{P}(A) \leqslant 1, \tag{2.3}$$

$$\mathbf{P}(A_1 - A_2) = \mathbf{P}(A_1) - \mathbf{P}(A_1 \cap A_2), \tag{2.4}$$

$$\mathbf{P}(A_2 - A_1) = \mathbf{P}(A_2) - \mathbf{P}(A_1 \cap A_2), \tag{2.5}$$

$$\mathbf{P}(A_1 \cup A_2) = \mathbf{P}(A_1) + \mathbf{P}(A_2) - \mathbf{P}(A_1 \cap A_2) \tag{2.6}$$

hold for arbitrary events A, A_1 and A_2. Moreover,

$$\mathbf{P}(A_1) \leqslant \mathbf{P}(A_2) \qquad \text{if} \quad A_1 \subset A_2. \tag{2.7}$$

Proof. Formula (2.3) follows at once from the interpretation of probability as the limiting value of relative frequency, since obviously

$$0 \leqslant \frac{n(A)}{n} \leqslant 1,$$

where $n(A)$ is the number of occurrences of an event A in n trials.[4] Given any two events A_1 and A_2, we have

$$A_1 = (A_1 - A_2) \cup (A_1 \cap A_2),$$
$$A_2 = (A_2 - A_1) \cup (A_1 \cap A_2),$$
$$A_1 \cup A_2 = (A_1 - A_2) \cup (A_2 - A_1) \cup (A_1 \cap A_2),$$

where the events $A_1 - A_2$, $A_2 - A_1$ and $A_1 \cap A_2$ are mutually exclusive. Therefore, by (2.2),

$$\mathbf{P}(A_1) = \mathbf{P}(A_1 - A_2) + \mathbf{P}(A_1 \cap A_2), \tag{2.8}$$

$$\mathbf{P}(A_2) = \mathbf{P}(A_2 - A_1) + \mathbf{P}(A_1 \cap A_2), \tag{2.9}$$

$$\mathbf{P}(A_1 \cup A_2) = \mathbf{P}(A_1 - A_2) + \mathbf{P}(A_2 - A_1) + \mathbf{P}(A_1 \cap A_2). \tag{2.10}$$

Formulas (2.8) and (2.9) are equivalent to (2.4) and (2.5). Then, using (2.4) and (2.5), we can write (2.10) in the form

$$\begin{aligned}\mathbf{P}(A_1 \cup A_2) &= \mathbf{P}(A_1) - \mathbf{P}(A_1 \cap A_2) + \mathbf{P}(A_2) \\ &\qquad - \mathbf{P}(A_1 \cap A_2) + \mathbf{P}(A_1 \cap A_2) \\ &= \mathbf{P}(A_1) + \mathbf{P}(A_2) - \mathbf{P}(A_1 \cap A_2),\end{aligned}$$

---

[4] Note that $\mathbf{P}(\varnothing) = 0$, $\mathbf{P}(\Omega) = 1$, since $n(\varnothing) = 0$, $n(\Omega) = n$ for all n. Thus the impossible event has probability zero, while the certain event has probability one.

thereby proving (2.6). Finally to prove (2.7), we note that if $A_1 \subset A_2$, then $A_1 \cap A_2 = A_1$, and hence (2.9) implies

$$\mathbf{P}(A_1) = \mathbf{P}(A_2) - \mathbf{P}(A_2 - A_1) \leqslant \mathbf{P}(A_2),$$

since $\mathbf{P}(A_2 - A_1) \geqslant 0$ by (2.3). ▮

The addition law (2.2) becomes much more complicated if we drop the requirement that the events be mutually exclusive:

THEOREM 2.2 *Given any n events $A_1, A_2, \ldots, A_n$, let*[5]

$$P_1 = \sum_{i=1}^{n} \mathbf{P}(A_i),$$

$$P_2 = \sum_{1 \leqslant i < j \leqslant n} \mathbf{P}(A_i A_j)$$

$$P_3 = \sum_{1 \leqslant i < j < k \leqslant n} \mathbf{P}(A_i A_j A_k), \ldots$$

Then

$$\mathbf{P}\left(\bigcup_{k=1}^{n} A_k\right) = P_1 - P_2 + P_3 - P_4 + \cdots \pm P_n. \qquad (2.11)$$

Proof. For $n = 2$, (2.11) reduces to formula (2.6), which we have already proved. Suppose (2.11) holds for any $n - 1$ events. Then

$$\mathbf{P}\left(\bigcup_{k=2}^{n} A_k\right) = \sum_{i=2}^{n} \mathbf{P}(A_i) - \sum_{2 \leqslant i < j \leqslant n} \mathbf{P}(A_i A_j) + \sum_{2 \leqslant i < j < k \leqslant n} \mathbf{P}(A_i A_j A_k) - \cdots \qquad (2.12)$$

and

$$\mathbf{P}\left(\bigcup_{k=2}^{n} A_1 A_k\right) = \sum_{i=2}^{n} \mathbf{P}(A_1 A_i) - \sum_{t \leqslant t < j \leqslant n} \mathbf{P}(A_1 A_i A_j) + \sum_{2 \leqslant i < j < k \leqslant n} \mathbf{P}(A_1 A_i A_j A_k) - \cdots. \qquad (2.13)$$

But, by (2.6),

$$\mathbf{P}\left(\bigcup_{k=1}^{n} A_k\right) = \mathbf{P}(A_1) + \mathbf{P}\left(\bigcup_{k=2}^{n} A_k\right) - \mathbf{P}\left(\bigcup_{k=2}^{n} A_1 A_k\right),$$

and hence, by (2.12) and (2.13),

$$\mathbf{P}\left(\bigcup_{k=1}^{n} A_k\right) = \mathbf{P}(A_1) + \sum_{i=2}^{n} \mathbf{P}(A_i) - \sum_{2 \leqslant i < j \leqslant n} \mathbf{P}(A_i A_j) + \sum_{2 \leqslant i < j < k \leqslant n} \mathbf{P}(A_i A_j A_k) - \cdots - \sum_{i=2}^{n} \mathbf{P}(A_1 A_i) + \sum_{2 \leqslant i < j \leqslant n} \mathbf{P}(A_1 A_j A_k) - \cdots = P_1 - P_2 + P_3 - \cdots,$$

---

[5] $A_i A_j$ is shorthand for the intersection $A_i \cap A_j$, $A_i A_j A_k$ is shorthand for $A_i \cap A_j \cap A_k$, and so on. In a sum like $\sum_{1 \leqslant i < j < k \leqslant n} \mathbf{P}(A_i A_j A_k)$, each group of indices (satisfying the indicated inequalities) is encountered just once.

i.e., (2.11) holds for any n events. The proof for all n now follows by mathematical induction. ∎

***Example* (*Coincidences*).** Suppose n students have n identical raincoats which they unwittingly hang on the same coat rack while attending class. After class, each student selects a raincoat at random, being unable to tell it apart from all the others. What is the probability that at least one raincoat ends up with its original owner?

Solution. We number both the students and the raincoats from 1 to n, with the kth raincoat belonging to the kth student ($k = 1, 2, \ldots, n$). Let A_k be the event that the kth student retrieves his own raincoat. Then the event A that "at least one raincoat ends up with its original owner" is just

$$A = \bigcup_{k=1}^{n} A_k.$$

Every outcome of the experiment consisting of "randomly selecting" the raincoats can be described by a permutation $(i_1, i_2, \ldots, i_n)$, where i_k is the number of the raincoat selected by the kth student. Consider the event $A_{k_1}A_{k_2}\cdots A_{k_m}$, where $m \leqslant n$. This event occurs whenever $i_{k_1} = k_1$, $i_{k_2} = k_2, \ldots, i_{k_m} = k_m$ and the other indices take the remaining $n - m$ values in any order. Therefore

$$\mathbf{P}(A_{k_1}A_{k_2}\cdots A_{k_m}) = \frac{N(A_{k_1}A_{k_2}\cdots A_{k_m})}{N} = \frac{(n-m)!}{n!},$$

where $N(A_{k_1}A_{k_2}\cdots A_{k_m}) = (n-m)!$ is just the total number of permutations of $n - m$ things, and $N = n!$ is the total number of permutations of n things (m is the number of fixed indices $k_1, k_2, \ldots, k_m$). There are precisely

$$C_m^n = \frac{n!}{m!\,(n-m)!}$$

distinct events of the type $A_{k_1}A_{k_2}\cdots A_{k_m}$, with m fixed indices, this being the number of combinations of n things taken m at a time (recall Theorem 1.3, p. 7). It follows that

$$P_m = \sum_{1 \leqslant k_1 < k_2 < \cdots < k_m \leqslant n} \mathbf{P}(A_{k_1}A_{k_2}\cdots A_{k_m}) = C_m^n \frac{(n-m)!}{n!} = \frac{1}{m!}$$

Hence, by formula (2.11),

$$\mathbf{P}\left(\bigcup_{k=1}^{n} A_k\right) = P_1 - P_2 + P_3 - P_4 + \cdots \pm P_n$$

$$= 1 - \frac{1}{2!} + \frac{1}{3!} - \frac{1}{4!} + \cdots \pm \frac{1}{n!},$$

i.e., the desired probability $\mathbf{P}(A)$ is a partial sum of the power series expansion of the function $1 - e^x$ with $x = -1$:

$$1 - e^{-1} = 1 - \frac{1}{2!} + \frac{1}{3!} - \frac{1}{4!} + \cdots \pm \frac{1}{n!} \pm \cdots.$$

Thus, for large n,

$$\mathbf{P}(A) \approx 1 - e^{-1} \approx 0.632. \tag{2.14}$$

To generalize the addition law to the case of an *infinite sequence* of mutually exclusive events $A_1, A_2, \ldots$, we repeatedly apply (2.1). Thus

$$\begin{aligned}\mathbf{P}(A_1 \cup A_2 \cup A_3 \cup \cdots) &= \mathbf{P}(A_1) + \mathbf{P}(A_2 \cup A_3 \cup \cdots)\\ &= \mathbf{P}(A_1) + \mathbf{P}(A_2) + \mathbf{P}(A_3 \cup \cdots)\\ &= \mathbf{P}(A_1) + \mathbf{P}(A_2) + \mathbf{P}(A_3) + \cdots,\end{aligned}$$

or equivalently,

$$\mathbf{P}\left(\bigcup_{k=1}^{\infty} A_k\right) = \sum_{k=1}^{\infty} \mathbf{P}(A_k).$$

We can combine this formula and (2.2) into a single formula

$$\mathbf{P}\left(\bigcup_{k} A_k\right) = \sum_{k} \mathbf{P}(A_k), \tag{2.2'}$$

where it will always be clear from the context whether $\bigcup_k$ and $\sum_k$ have finite or infinite limits.[6]

The "generalized addition law" (2.2′) has a number of important consequences. We begin with two theorems expressing a kind of "continuity property" of probability:

THEOREM 2.3. *If $A_1, A_2, \ldots$ is an "increasing sequence" of events, i.e., a sequence such that $A_1 \subset A_2 \subset \cdots$, then*

$$\mathbf{P}\left(\bigcup_{k} A_k\right) = \lim_{n \to \infty} \mathbf{P}(A_n). \tag{2.15}$$

Proof. Clearly, the events

$$B_1 = A_1, \quad B_2 = A_2 - A_1, \ldots, \quad B_n = A_n - \bigcup_{k=1}^{n-1} B_k, \ldots \tag{2.16}$$

[6] In the last analysis, formulas (2.1), (2.2′) and (2.3) are *axioms*, although they are, of course, strongly suggested by experience, i.e., by the interpretation of probabilities as limiting values of relative frequencies. In this sense, they are the "only reasonable axioms," and lead to a model of random phenomena whose consequences are fully confirmed by experiment.

are mutually exclusive and have union $\bigcup_k A_k$. Moreover,

$$\bigcup_{k=1}^{n} B_k = A_n.$$

Therefore, by (2.2′),

$$\mathbf{P}\left(\bigcup_k A_k\right) = \mathbf{P}\left(\bigcup_k B_k\right) = \sum_k \mathbf{P}(B_k) = \lim_{n\to\infty} \sum_{k=1}^{n} \mathbf{P}(B_k)$$
$$= \lim_{n\to\infty} \mathbf{P}\left(\bigcup_{k=1}^{n} B_k\right) = \lim_{n\to\infty} \mathbf{P}(A_n). \quad \blacksquare$$

Similarly, we have

THEOREM 2.3′. *If $A_1, A_2, \ldots$ is a "decreasing sequence" of events, i.e., a sequence such that $A_1 \supset A_2 \supset \cdots$, then*

$$\mathbf{P}\left(\bigcap_k A_k\right) = \lim_{n\to\infty} \mathbf{P}(A_n).$$

Proof. Going over to complementary events, we have $\bar{A}_1 \subset \bar{A}_2 \subset \cdots$, and hence, by (2.15),

$$\mathbf{P}\left(\bigcap_k A_k\right) = 1 - \mathbf{P}\left(\bigcup_k \bar{A}_k\right) = 1 - \lim_{n\to\infty} \mathbf{P}(\bar{A}_n)$$
$$= \lim_{n\to\infty} [1 - \mathbf{P}(\bar{A}_n)] = \lim_{n\to\infty} \mathbf{P}(A_n). \quad \blacksquare$$

In the case of arbitrary events, we must replace $=$ by $\leqslant$ in (2.2′):

THEOREM 2.4. *The ineqaulity*

$$\mathbf{P}\left(\bigcup_k A_k\right) \leqslant \sum_k \mathbf{P}(A_k)$$

holds for arbitrary events $A_1, A_2, \ldots$.

Proof. As in the proof of Theorem 2.3, $\bigcup_k A_k$ is the union of the mutually exclusive events (2.16), where obviously $B_k \subset A_k$ and hence $\mathbf{P}(B_k) \leqslant \mathbf{P}(A_k)$, by (2.7). Therefore

$$\mathbf{P}\left(\bigcup_k A_k\right) = \mathbf{P}\left(\bigcup_k B_k\right) = \sum_k \mathbf{P}(B_k) \leqslant \sum_k \mathbf{P}(A_k). \quad \blacksquare$$

Finally, we prove a proposition that will be needed in Chapter 7:

THEOREM 2.5 ***(First Borel-Cantelli lemma)***.[7] *Given a sequence of events*

[7] For the "second Borel-Cantelli lemma," see Theorem 3.1, p. 33.

$A_1, A_2, \ldots,$ *with probabilities* $p_k = \mathbf{P}(A_k)$, $k = 1, 2, \ldots,$ *suppose*

$$\sum_{k=1}^{\infty} p_k < \infty, \tag{2.17}$$

i.e., suppose the series on the left converges. Then, with probability 1 *only finitely many of the events* $A_1, A_2, \ldots$, *occur.*

Proof. Let B be the event that infinitely many of the events A_1, $A_2, \ldots$ occur, and let

$$B_n = \bigcup_{k \geqslant n} A_k,$$

so that B_n is the event that at least one of the events $A_n, A_{n+1}, \ldots$ occurs. Clearly B occurs if and only if B_n occurs for every $n = 1$, $2, \ldots$ Therefore

$$B = \bigcap_n B_n = \bigcap_n \left(\bigcup_{k \geqslant n} A_k \right).$$

Moreover, $B_1 \supset B_2 \supset \cdots$, and hence, by Theorem 2.3′,

$$\mathbf{P}(B) = \lim_{n \to \infty} \mathbf{P}(B_n).$$

But, by Theorem 2.4,

$$\mathbf{P}(B_n) \leqslant \sum_{k \geqslant n} \mathbf{P}(A_k) = \sum_{k \geqslant n} p_k \to 0 \text{ as } n \to \infty,$$

because of (2.17). Therefore

$$\mathbf{P}(B) = \lim_{n \to \infty} \mathbf{P}(B_n) = 0,$$

i.e., the probability of infinitely many of the events $A_1, A_2, \ldots$ occurring is 0. Equivalently, the probability of only finitely many of the events $A_1, A_2, \ldots$ occurring is 1. ▌

PROBLEMS

1. Interpret the following relations involving events A, B and C:
a) $AB = A$; b) $ABC = A$; c) $A \cup B \cup C = A$.

2. When do the following relations involving the events A and B hold:
a) $A \cup B = \bar{A}$; b) $AB = \bar{A}$; c) $A \cup B = AB$?

3. Simplify the following expressions involving events A, B and C:
a) $(A \cup B)(B \cup C)$; b) $(A \cup B)(A \cup \bar{B})$; c) $(A \cup B)(A \cup \bar{B})(\bar{A} \cup B)$.
Ans. a) $AC \cup B$; b) A; c) AB.

4. Given two events A and B, find the event X such that

$$\overline{(X \cup A)} \cup \overline{(X \cup \bar{A})} = B.$$

Ans. $X = \bar{B}$.

5. Let A be the event that at least one of three inspected items is defective, and B the event that all three items are of acceptable quality. What are the events $A \cup B$ and AB?

6. A whole number from 1 to 1000 is chosen at random. Let A be the event that the number is divisible by 5, and B the event that the number ends in a zero. What is the event $A\bar{B}$?

7. A target is made up of 10 circular disks bounded by 10 concentric circles of radii $r_1, r_2, \ldots, r_{10}$ where $r_1 < r_2 < \cdots < r_{10}$. Let A_k be the event consisting of the disk of radius r_k being hit ($k = 1, 2, \ldots, 10$). What are the events

$$B = \bigcup_{k=1}^{6} A_k, \qquad C = \bigcap_{k=5}^{10} A_k?$$

Ans. $B = A_6$, $C = A_5$.

8. Given any event A, prove that

$$\mathbf{P}(A) = 1 - \mathbf{P}(\bar{A}), \qquad \mathbf{P}(\bar{A}) = 1 - \mathbf{P}(A).$$

9. A marksman fires at a target made up of a central circular disk and two concentric rings. The probabilities of hitting the disk and the rings are 0.35, 0.30 and 0.25, respectively. What is the probability of missing the target?

10. Five items are chosen at random from a batch of 100 items and then inspected. The whole batch is rejected if any of the items is found to be defective. What is the probability of the batch being rejected if it contains 5 defective items?

Ans. $1 - \dfrac{95 \cdot 94 \cdot 93 \cdot 92 \cdot 91}{100 \cdot 99 \cdot 98 \cdot 97 \cdot 96} \approx 0.23.$

11. A secretary forgets the last digit of a telephone number, and dials the last digit at random. What is the probability of calling no more than three wrong numbers? How is this probability changed if she recalls that the last digit is even?

12. Given any n events $A_1, A_2, \ldots, A_n$, prove that

$$\mathbf{P}\left(\bigcap_{k=1}^{n} A_k\right) = 1 - \mathbf{P}\left(\bigcup_{k=1}^{n} \bar{A}_k\right).$$

13. A batch of 100 manufactured items contains 5 defective items. Fifty items are chosen at random and then inspected. Suppose the whole batch is accepted if no more than one of the 50 inspected items is defective. What is the probability of accepting the whole batch?

Ans. $\dfrac{47 \cdot 37}{99 \cdot 97} \approx 0.18.$

14. Write an expression for the probability $p(r)$ that among r randomly selected people, at least two have a common birthday.

Comment. Rather surprisingly, it turns out that $p(r) > \frac{1}{2}$ if $r = 23$.[8]

---

[8] See W. Feller, *op. cit.*, p. 33.

15. Test the approximation (2.14) for $n = 3, 4, 5$ and 6.

16. Use Theorem 2.2 and Stirling's formula to find the probability that some player is dealt a complete suit in a game of bridge.

Ans. $\dfrac{16}{C_{13}^{52}} - \dfrac{72}{C_{13}^{52}C_{13}^{39}} + \dfrac{72}{C_{13}^{52}C_{13}^{39}C_{13}^{26}} \approx \dfrac{3^{39}}{4^{50}}\sqrt{\dfrac{39\pi}{2}} \approx \frac{1}{4} \times 10^{-10}.$

17. Given any n events $A_1, A_2, \ldots, A_n$, prove that the probability of exactly m $(m \leqslant n)$ of the events occurring is

$$P_m - \binom{m+1}{m}P_{m+1} + \binom{m+2}{m}P_{m+2} - \cdots \pm \binom{n}{m}P_n,$$

where $P_m, P_{m+1}, \ldots$ are the same as in Theorem 2.2.

18. Let $n = 10$ in the example on p. 19. What is the probability that exactly 5 raincoats end up with their original owners?

19. A whole number from 1 to 1000 is chosen at random. What is the probability of its being a power (higher than the first) of another whole number?

Hint. $31^2 < 1000 < 32^2$.

Ans. $\frac{1}{25}$.

3

DEPENDENT EVENTS

5. Conditional Probability

In observing the outcomes of a random experiment, one is often interested in how the outcome of one event A is influenced by that of another event B. For example, in one extreme case the relation between A and B may be such that A always occurs if B does, while in the other extreme case A never occurs if B does. To characterize the relation between A and B, we introduce the *conditional probability of A on the hypothesis B*, i.e., the "probability of A occurring under the condition that B is known to have occurred." This quantity is defined by

$$\mathbf{P}(A \mid B) = \frac{\mathbf{P}(AB)}{\mathbf{P}(B)}, \tag{3.1}$$

where AB is the intersection of the events A and B, and it is assumed that $\mathbf{P}(B) > 0$.

To clarify the meaning of (3.1), consider an experiment with a finite number of equiprobable outcomes (elementary events). Let N be the total number of outcomes, $N(B)$ the number of outcomes leading to the occurrence of the event B, and $N(AB)$ the number of outcomes leading to the occurrence of both A and B. Then, as on p. 1, the probabilities of B and AB are just

$$\mathbf{P}(B) = \frac{N(B)}{N}, \qquad \mathbf{P}(AB) = \frac{N(AB)}{N}, \tag{3.2}$$

and hence (3.1) implies

$$\mathbf{P}(A \mid B) = \frac{N(AB)}{N(B)}. \tag{3.3}$$

But (3.3) is of the same form as (3.2), if we restrict the set of possible outcomes to those in which B is known to have occurred. In fact, the denominator in (3.3) is the total number of such outcomes, while the numerator is the total number of such outcomes leading to the occurrence of A.

It is easy to see that conditional probabilities have properties analogous to those of ordinary probabilities. For example,

a) $0 \leqslant \mathbf{P}(A \mid B) \leqslant 1$;
b) If A and B are incompatible, so that $AB = \varnothing$, then $\mathbf{P}(A \mid B) = 0$;
c) If B implies A, so that $B \subset A$, then $\mathbf{P}(A \mid B) = 1$;
d) If $A_1, A_2, \ldots$ are mutually exclusive events with union $A = \bigcup_k A_k$, then

$$\mathbf{P}(A \mid B) = \sum_k \mathbf{P}(A_k \mid B) \tag{3.4}$$

(the addition law for conditional probabilities).

Property a) is an immediate consequence of (3.1) and the formula $0 \leqslant \mathbf{P}(AB) \leqslant \mathbf{P}(B)$, implied by $\varnothing \subset AB \subset B$. To prove b), we note that $AB = \varnothing$ implies $\mathbf{P}(AB) = 0$ and hence $\mathbf{P}(A \mid B) = 0$, by (3.1). Similarly, c) follows from the observation that if $B \subset A$, then $AB = B$, $\mathbf{P}(AB) = \mathbf{P}(B)$, and hence $\mathbf{P}(A \mid B) = 1$, by (3.1). Finally, if $A = \bigcup_k A_k$, where $A_1, A_2, \ldots$ are mutually exclusive events, then

$$AB = \bigcup_k A_k B,$$

and hence

$$\mathbf{P}(AB) = \sum_k \mathbf{P}(A_k B), \tag{3.5}$$

by formula (2.2′), p. 20, the addition law for ordinary probabilities. Dividing (3.5) by $\mathbf{P}(B)$, we get (3.4), because of (3.1) and

$$\mathbf{P}(A_k \mid B) = \frac{\mathbf{P}(A_k B)}{\mathbf{P}(B)}.$$

In calculating the probability of an event A, it is often convenient to use conditional probabilities as an intermediate step. Suppose $B_1, B_2, \ldots$ is a "full set"[1] of mutually exclusive events, in the sense that one (and only one) of the events $B_1, B_2, \ldots$ always occurs. Then we can find $\mathbf{P}(A)$ by using the "total probability formula"

$$\mathbf{P}(A) = \sum_k \mathbf{P}(A \mid B_k)\mathbf{P}(B_k). \tag{3.6}$$

---

[1] Synonymously, an "exhaustive set."

To prove (3.6), we need only note that

$$\bigcup_k B_k = \Omega,$$

where Ω is the whole sample space, since one of the events $B_1, B_2, \ldots$ must occur. But then

$$A = \bigcup_k AB_k,$$

and hence

$$\mathbf{P}(A) = \mathbf{P}\left(\bigcup_k AB_k\right) = \sum_k \mathbf{P}(AB_k) = \sum_k \frac{\mathbf{P}(AB_k)}{\mathbf{P}(B_k)} \mathbf{P}(B_k),$$

which is equivalent to (3.6).

***Example* 1.** A hiker leaves the point O shown in Figure 3, choosing one of the roads OB_1, OB_2, OB_3, OB_4 at random. At each subsequent crossroads he again chooses a road at random. What is the probability of the hiker arriving at the point A?

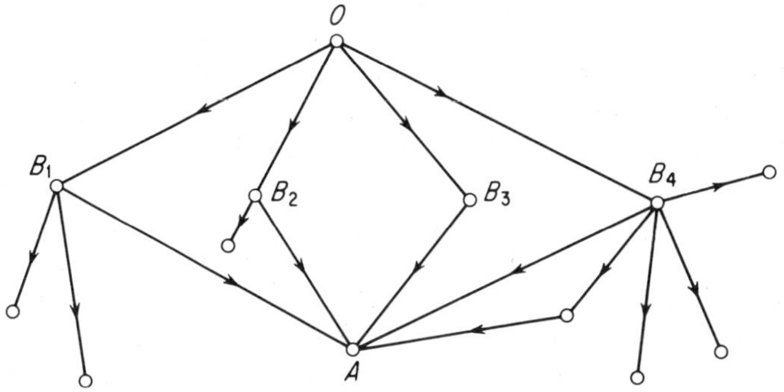

FIGURE 3

Solution. Let the event that the hiker passes through the point B_k, $k = 1, \ldots, 4$, be denoted by the same symbol B_k as the point itself. Then B_1, B_2, B_3, B_4 form a "full set" of mutually exclusive events, since the hiker must pass through one of these points. Moreover, the events B_1, B_2, B_3, B_4 are equiprobable, since, by hypothesis, the hiker initially makes a completely random choice of one of the roads OB_1, OB_2, OB_3, OB_4. Therefore

$$\mathbf{P}(B_k) = \frac{1}{4}, \qquad k = 1, \ldots, 4.$$

Once having arrived at B_1, the hiker can proceed to A only by making the proper choice of one of three equiprobable roads. Hence the conditional probability of arriving at A starting from B_1 is just $\frac{1}{3}$. Let the event that the hiker arrives at A be denoted by the same symbol A as the point itself. Then

$$\mathbf{P}(A \mid B_1) = \frac{1}{3}.$$

and similarly

$$\mathbf{P}(A \mid B_2) = \frac{1}{2}, \quad \mathbf{P}(A \mid B_3) = 1, \quad \mathbf{P}(A \mid B_4) = \frac{2}{5}$$

(consult the figure). It follows from (3.6) that the probability of arriving at A is

$$\begin{aligned}\mathbf{P}(A) &= \mathbf{P}(A \mid B_1)\mathbf{P}(B_1) + \mathbf{P}(A \mid B_2)\mathbf{P}(B_2) \\ &\quad + \mathbf{P}(A \mid B_3)\mathbf{P}(B_3) + \mathbf{P}(A \mid B_4)\mathbf{P}(B_4) \\ &= \frac{1}{4}\left(\frac{1}{3} + \frac{1}{2} + 1 + \frac{2}{5}\right) = \frac{67}{120}.\end{aligned}$$

Example 2 (The optimal choice problem). Consider a set of m objects, all of different quality, such that it is always possible to tell which of a given pair of objects is better. Suppose the objects are presented one at a time and at random to an observer, who at each stage either selects the object, thereby designating it as "the best" and examining no more objects, or rejects the object *once and for all* and examines another one. (Of course, the observer may very well make the mistake of rejecting the best object in the vain hope of finding a better one!) For example, the observer may be a fussy young lady and the objects a succession of m suitors. At each stage, she can either accept the suitor's proposal of marriage, thereby terminating the process of selecting a husband, or she may reject him (thereby losing him forever) and wait for a better prospect to come along. It will further be assumed that the observer adopts the following natural rule for selecting the best object: "Never select an object inferior to those previously rejected." Then the observer can select the first object and stop looking for a better one, or he can reject the first object and examine further objects one at a time until he finds one better than those previously examined. He can then select this object, thereby terminating the inspection process, or he can examine further objects in the hope of eventually finding a still better one, and so on. Of course, it is entirely possible that he will reject the very best object somewhere along the line, and hence never be able to make a selection at all. On the other hand, if the number of objects is large, almost anyone would reject the first object in the hope of eventually finding a better one.

Now suppose the observer, following the above "decision rule," selects the ith inspected object once and for all, giving up further inspection. (The ith object must then be better than the $i - 1$ previously inspected objects.) What is the probability that this ith object is actually the best of all m objects, both inspected and uninspected?

Solution. Let B be the event that the last of the i inspected objects is the best of those inspected, and let A be the event that the ith object is the best of all m objects, both inspected and uninspected. Then we want the conditional probability $\mathbf{P}(A \mid B)$ of the event A given that B has already occurred. According to (3.1), to calculate $\mathbf{P}(A \mid B)$ we need both $\mathbf{P}(B)$ and $\mathbf{P}(AB)$. Obviously $A \subset B$ and hence $AB = A$, so that $\mathbf{P}(AB) = \mathbf{P}(A)$. By hypothesis, all possible arrangements of the objects in order of presentation are equiprobable (the objects are presented "at random"). Hence $\mathbf{P}(B)$ is the probability that in a random permutation of i distinguishable objects (the objects differ in quality) a given object (the best of all i objects) occupies the ith place. Since there are $i!$ permutations of all i objects and $(i - 1)!$ permutations subject to the condition that a given object occupy the ith place, this probability is just

$$\mathbf{P}(B) = \frac{(i-1)!}{i!} = \frac{1}{i}.$$

Similarly, $\mathbf{P}(A)$ is the probability that in a random permutation of m distinguishable objects, a given object (the best of all m objects) occupies the ith place, and hence

$$\mathbf{P}(A) = \frac{(m-1)!}{m!} = \frac{1}{m}.$$

Therefore the desired conditional probability $\mathbf{P}(A \mid B)$ is just

$$\mathbf{P}(A \mid B) = \frac{\mathbf{P}(AB)}{\mathbf{P}(B)} = \frac{\mathbf{P}(A)}{\mathbf{P}(B)} = \frac{i}{m}.$$

***Example* 3 (*The gambler's ruin*).** Consider the game of "heads or tails," in which a coin is tossed and a player wins 1 dollar, say, if he successfully calls the side of the coin which lands upward, but otherwise loses 1 dollar. Suppose the player's initial capital is x dollars, and he intends to play until he wins m dollars but no longer. In other words, suppose the game continues until the player either wins the amount of m dollars, stipulated in advance, or else loses all his capital and is "ruined." What is the probability that the player will be ruined?

Solution. The probability of ruin clearly depends on both the initial capital x and the final amount m. Let $p(x)$ be the probability of the player's being ruined if he starts with a capital of x dollars. Then the probability of ruin given that the player wins the first call is just $p(x + 1)$, since the player's capital becomes $x + 1$ if he wins the first call. Similarly, the probability of ruin given that the player loses the first call is $p(x - 1)$, since the player's capital becomes $x - 1$ if he loses the first call. In other words, if B_1 is the event that the player wins the first call and B_2 the event that he loses the first call, while A is the event of ruin, then

$$\mathbf{P}(A \mid B_1) = p(x+1), \qquad \mathbf{P}(A \mid B_2) = p(x-1).$$

The mutually exclusive events B_1 and B_2 form a "full set," since the player either wins or loses the first call. Moreover, we have

$$\mathbf{P}(B_1) = \frac{1}{2}, \qquad \mathbf{P}(B_2) = \frac{1}{2},$$

assuming fair tosses of an unbiased coin (cf. Problem 1, p. 65). Hence, by (3.6),

$$\mathbf{P}(A) = \mathbf{P}(A \mid B_1)\mathbf{P}(B_1) + \mathbf{P}(A \mid B_2)\mathbf{P}(B_2),$$

i.e.,

$$p(x) = \frac{1}{2}\,[p(x+1) + p(x-1)], \qquad 1 \leqslant x \leqslant m-1, \tag{3.7}$$

where obviously

$$p(0) = 1, \qquad p(m) = 0. \tag{3.8}$$

The solution of (3.7) is the linear function

$$p(x) = C_1 + C_2 x, \tag{3.9}$$

where the coefficients C_1 and C_2 are determined by the boundary conditions (3.8), which imply

$$C_1 = 1, \qquad C_1 + C_2 m = 0. \tag{3.10}$$

Combining (3.9) and (3.10), we finally find that the probability of ruin given an initial capital of x dollars is just

$$p(x) = 1 - \frac{x}{m}, \qquad 0 \leqslant x \leqslant m.$$

6. Statistical Independence

In saying that two experiments are "statistically independent" (or briefly, "independent"), we mean, roughly speaking, that the outcome of one experiment has no influence on the outcome of the other. Let A_1 be an event associated only with the first experiment, and A_2 an event associated only with the second experiment. Then the occurrence of A_1 has no influence on the probability of occurrence of A_2, and conversely. In this sense, we say that the events A_1 and A_2 are "(statistically) independent."

To give mathematical expression to the notion of independence, we calculate the probability that two independent events A_1 and A_2 both occur.

To this end, we again resort to the empirical fact that the relative frequency of an event in a large series of "independent trials under identical conditions"[2] virtually coincides with its probability (recall Sec. 1). Imagine a long series of such trials, where each trial involves carrying out *both* experiments. If n is the total number of trials and $n(A_1A_2)$ the number of trials leading to occurrence of both A_1 and A_2, then

$$\mathbf{P}(A_1A_2) \sim \frac{n(A_1A_2)}{n}. \tag{3.11}$$

Moreover, if $n(A_2)$ is the number of trials leading to occurrence of A_2, then

$$\mathbf{P}(A_2) \sim \frac{n(A_2)}{n}. \tag{3.12}$$

Suppose we confine ourselves to examining the results of the $n(A_2)$ trials leading to occurrence of A_2, and look for occurrence of A_1. Then clearly A_1 will occur in precisely $n(A_1A_2)$ of these trials. Moreover, if n is very large, then so is $n(A_2)$, and hence

$$\mathbf{P}(A_1) \sim \frac{n(A_1A_2)}{n(A_2)}, \tag{3.13}$$

since A_2 is associated only with the second experiment, which has nothing whatsoever to do with the first experiment or the event A_1 associated with it. Combining (3.11)–(3.13), we find that

$$\mathbf{P}(A_1A_2) \sim \frac{n(A_1A_2)}{n} = \frac{n(A_1A_2)}{n(A_2)}\frac{n(A_2)}{n} \sim \mathbf{P}(A_1)\mathbf{P}(A_2),$$

or, after going over to exact equations (in the limit as $n \to \infty$),

$$\mathbf{P}(A_1A_2) = \mathbf{P}(A_1)\mathbf{P}(A_2). \tag{3.14}$$

Two events A_1 and A_2 are said to be (*statistically*) *independent* if they satisfy (3.14) and (*statistically*) *dependent* otherwise.[3]

The definition (3.14) is in keeping with the notion of conditional probability introduced in Sec. 5. In fact, if two events A_1 and A_2 are independent, then, loosely speaking, the occurrence of A_2 should have no influence on the probability of occurrence of A_1, and hence the conditional probability

---

[2] Thus there remains the problem of just what is meant by "independent trials under identical conditions" (a phrase already encountered on pp. 2 and 16), although the intuitive meaning of the phrase is perfectly clear, e.g., in a series of coin tosses. For a rigorous discussion of this whole issue, see W. Feller, *op. cit.*, p. 128.

[3] In the last analysis, (3.14) is a *definition*, although one strongly suggested by experience, i.e., by the intuitive meaning of independence and the interpretation of probabilities as limiting values of relative frequencies (recall footnote 6, p. 20).

$\mathbf{P}(A_1 \mid A_2)$ of A_1 occurring given that A_2 has already occurred should be the same as the unconditional probability of A_1, i.e.,

$$\mathbf{P}(A_1 \mid A_2) = \mathbf{P}(A_1)$$

(and similarly with A_1 and A_2 changing places). But clearly

$$\mathbf{P}(A_1 \mid A_2) = \frac{\mathbf{P}(A_1A_2)}{\mathbf{P}(A_2)} = \mathbf{P}(A_1)$$

if and only if (3.14) holds.

***Example* 1.** Let A_1 be the event that a card picked at random from a full deck is a spade, and A_2 the event that it is a queen. Are A_1 and A_2 independent events?

Solution. The question is not easily answered on the basis of physical intuition alone. However, noting that a full deck (52 cards) contains 13 spades and 4 queens, but only one queen of spades, we see at once that

$$\mathbf{P}(A_1) = \frac{13}{52} = \frac{1}{4}, \quad \mathbf{P}(A_2) = \frac{4}{52} = \frac{1}{13}, \quad \mathbf{P}(A_1A_2) = \frac{1}{52},$$

and hence $\mathbf{P}(A_1A_2) = \mathbf{P}(A_1)\mathbf{P}(A_2)$. Therefore the events A_1 and A_2 are independent.

***Example* 2.** In throwing a pair of dice, let A_1 be the event that "the first die turns up odd," A_2 the event that "the second die turns up odd," and A_3 the event that "the total number of spots is odd." Clearly, the number of spots on one die has nothing to do with the number of spots on the other die, and hence the events A_1 and A_2 are independent, with probabilities

$$\mathbf{P}(A_1) = \frac{1}{2}, \qquad \mathbf{P}(A_2) = \frac{1}{2}.$$

Moreover, it is clear that

$$\mathbf{P}(A_3) = \frac{1}{2}.$$

Given that A_1 has occurred, A_3 can occur only if the second die turns up even. Hence

$$\mathbf{P}(A_3 \mid A_1) = \frac{1}{2},$$

and similarly

$$\mathbf{P}(A_3 \mid A_2) = \frac{1}{2}.$$

It follows that

$$\mathbf{P}(A_3 \mid A_1) = \mathbf{P}(A_3), \quad \mathbf{P}(A_3 \mid A_2) = \mathbf{P}(A_3).$$

Therefore the events A_1 and A_3 are independent, and so are the events A_2 and A_3.

Generalizing (3.14), we have the following

DEFINITION. *The events $A_1, A_2, \ldots, A_n$ are said to be* (***mutually***) ***independent*** *if*

$$\mathbf{P}(A_iA_j) = \mathbf{P}(A_i)\mathbf{P}(A_j),$$

$$\mathbf{P}(A_iA_jA_k) = \mathbf{P}(A_i)\mathbf{P}(A_j)\mathbf{P}(A_k),$$

$$\cdots\cdots\cdots\cdots\cdots\cdots\cdots\cdots$$

$$\mathbf{P}(A_1A_2\cdots A_n) = \mathbf{P}(A_1)\mathbf{P}(A_2)\cdots\mathbf{P}(A_n)$$

for all combinations of indices such that $1 \leqslant i < j < \cdots < k \leqslant n$.

***Example* 3.** The events A_1, A_2 and A_3 in Example 2 are not independent, even though they are "pairwise independent" in the sense that

$$\mathbf{P}(A_iA_j) = \mathbf{P}(A_i)\mathbf{P}(A_j)$$

for all $1 \leqslant i < j \leqslant 3$. In fact, A_3 obviously cannot occur if A_1 and A_2 both occur, and hence

$$\mathbf{P}(A_1A_2A_3) = 0.$$

But

$$\mathbf{P}(A_1)\mathbf{P}(A_2)\mathbf{P}(A_3) = \frac{1}{2}\cdot\frac{1}{2}\cdot\frac{1}{2} = \frac{1}{8},$$

so that

$$\mathbf{P}(A_1A_2A_3) \neq \mathbf{P}(A_1)\mathbf{P}(A_2)\mathbf{P}(A_3).$$

Given an infinite sequence of events $A_1, A_2, \ldots$, suppose the events $A_1, \ldots, A_n$ are independent for every n. Then $A_1, A_2, \ldots$ is said to be a *sequence of independent events.*

THEOREM 3.1 (***Second Borel-Cantelli lemma***). *Given a sequence of independent events $A_1, A_2, \ldots$, with probabilities $p_k = \mathbf{P}(A_k)$, $k = 1, 2, \ldots$, suppose*

$$\sum_{k=1}^{\infty} p_k = \infty, \tag{3.15}$$

i.e., suppose the series on the left diverges. Then, with probability 1 *infinitely many of the events $A_1, A_2, \ldots$ occur.*

Proof. As in the proof of the first Borel-Cantelli lemma (Theorem 2.5, p. 21), let

$$B_n = \bigcup_{k \geqslant n} A_k, \qquad B = \bigcap_n B_n = \bigcap_n \left(\bigcup_{k \geqslant n} A_k \right),$$

so that B occurs if and only if infinitely many of the events $A_1, A_2, \ldots$ occur. Taking complements, we have

$$\bar{B}_n = \bigcap_{k \geqslant n} \bar{A}_k, \qquad \bar{B} = \bigcup_n \bar{B}_n.$$

Clearly,

$$\bar{B}_n \subset \bigcap_{k=n}^{n+m} \bar{A}_k$$

for every $m = 0, 1, 2, \ldots$ Therefore

$$\mathbf{P}(\bar{B}_n) \leqslant \mathbf{P}\left(\bigcap_{k=n}^{n+m} \bar{A}_k \right) = \mathbf{P}(\bar{A}_n) \cdots \mathbf{P}(\bar{A}_{n+m})$$

$$= (1 - p_n) \cdots (1 - p_{n+m}) \leqslant \exp\left(-\sum_{k=n}^{n+m} p_k \right), \tag{3.16}$$

where we use the inequality $1 - x \leqslant e^{-x}$, $x \geqslant 0$ and the fact that if A_1, $A_2, \ldots$ is a sequence of independent events, then so is the sequence of complementary events $\bar{A}_1, \bar{A}_2, \ldots$[4] But

$$\sum_{k=n}^{n+m} p_k \to \infty \quad \text{as} \quad m \to \infty,$$

because of (3.15). Therefore, passing to the limit $m \to \infty$ in (3.16), we find that $\mathbf{P}(\bar{B}_n) = 0$ for every $n = 1, 2, \ldots$ It follows that

$$\mathbf{P}(\bar{B}) \leqslant \sum_n \mathbf{P}(\bar{B}_n) = 0,$$

and hence

$$\mathbf{P}(B) = 1 - \mathbf{P}(\bar{B}) = 1,$$

i.e., the probability of infinitely many of the events $A_1, A_2, \ldots$ occurring is 1. ▌

PROBLEMS

1. Given any events A and B, prove that the events A, $\bar{A}B$ and $\overline{A \cup B}$ form a full set of mutually exclusive events.

---

[4] It is intuitively clear that if the events $A_1, \ldots, A_n$ are independent, then so are their complements. Concerning the rigorous proof of this fact, see Problem 7 and W. Feller, *op. cit.*, pp. 126, 128.

2. In a game of chess, let A be the event that White wins and B the event that Black wins. What is the event C such that A, B and C form a full set of mutually exclusive events?

3. Prove that if $\mathbf{P}(A \mid B) > \mathbf{P}(A)$, then $\mathbf{P}(B \mid A) > \mathbf{P}(B)$.

4. Prove that if $\mathbf{P}(A) = \mathbf{P}(B) = \frac{2}{3}$, then $\mathbf{P}(A \mid B) \geqslant \frac{1}{2}$.

5. Given any three events A, B and C, prove that

$$\mathbf{P}(ABC) = \mathbf{P}(A)\mathbf{P}(B \mid A)\mathbf{P}(C \mid AB).$$

Generalize this formula to the case of any n events.

6. Verify that

$$\mathbf{P}(A) = \mathbf{P}(A \mid B) + \mathbf{P}(A \mid \bar{B})$$

if

a) $A = \varnothing$; b) $B = \varnothing$; c) $B = \Omega$; d) $B = A$; e) $B = \bar{A}$.

7. Prove that if the events A and B are independent, then so are their complements.

Hint. Clearly $\mathbf{P}(B \mid A) + \mathbf{P}(\bar{B} \mid A) = 1$ for arbitrary A and B. Moreover $\mathbf{P}(B \mid A) = \mathbf{P}(B)$, by hypothesis. Therefore $\mathbf{P}(\bar{B} \mid A) = 1 - \mathbf{P}(B) = \mathbf{P}(\bar{B})$, so that A and $\bar{B}$ are independent.

8. Two events A and B with positive probabilities are incompatible. Are they dependent?

9. Consider n urns, each containing w white balls and b black balls. A ball is drawn at random from the first urn and put into the second urn, then a ball is drawn at random from the second urn and put into the third urn, and so on, until finally a ball is drawn from the last urn and examined. What is the probability of this ball being white?

Ans. $\dfrac{w}{w+b}$.

10. In Example 1, p. 27, find the probability of the hiker arriving at each of the 6 destinations other than A. Verify that the sum of the probabilities of arriving at all possible destinations is 1.

11. Prove that the probability of ruin in Example 3, p. 29 does not change if the stakes are changed.

12. Prove that the events A and B are independent if $\mathbf{P}(B \mid A) = \mathbf{P}(B \mid \bar{A})$.

13. One urn contains w_1 white balls and b_1 black balls, while another urn contains w_2 white balls and b_2 black balls. A ball is drawn at random from each urn, and then one of the two balls so obtained is chosen at random. What is the probability of this ball being white?

Ans. $\dfrac{1}{2}\left(\dfrac{w_1}{w_1+b_1} + \dfrac{w_2}{w_2+b_2}\right)$.

14. Nine out of 10 urns contain 2 white balls and 2 black balls each, while the other urn contains 5 white balls and 1 black ball. A ball drawn from a randomly chosen urn turns out to be white. What is the probability that the ball came from the urn containing 5 white balls?

Hint. If $B_1, \ldots, B_n$ is a full set of mutually exclusive events, then

$$\mathbf{P}(B_k \mid A) = \frac{\mathbf{P}(B_k)\mathbf{P}(A \mid B_k)}{\mathbf{P}(A)} = \frac{\mathbf{P}(B_k)\mathbf{P}(A \mid B_k)}{\sum_{k=1}^{n} \mathbf{P}(B_k)\mathbf{P}(A \mid B_k)},$$

a formula known as *Bayes' rule*. The events $B_1, \ldots, B_n$ are often regarded as "hypotheses" accounting for the occurrence of A.

Ans. $\frac{5}{32}$.

15. One urn contains only white balls, while another urn contains 30 white balls and 10 black balls. An urn is selected at random, and then a ball is drawn (at random) from the urn. The ball turns out to be white, and is then put back into the urn. What is the probability that another ball drawn from the same urn will be black?

Ans. $\frac{3}{28}$.

16. Two balls are drawn from an urn containing n balls numbered from 1 to n. The first ball is kept if it is numbered 1, and returned to the urn otherwise. What is the probability of the second ball being numbered 2?

Ans. $\dfrac{n^2 - n + 1}{n^2(n - 1)}$.

17. A regular tetrahedron is made into an unbiased die, by labelling the four faces a, b, c and abc, respectively. Let A be the event that the die falls on either of the two faces bearing the letter a, B the event that it falls on either of the two faces bearing the letter b, and C the event that it falls on either of the two faces bearing the letter c. Prove that the events A, B and C are "pairwise independent"[5] but not independent.

18. An urn contains w white balls, b black balls and r red balls. Find the probability of a white ball being drawn before a black ball if

a) Each ball is replaced after being drawn;
b) No balls are replaced.

Ans. $\dfrac{w}{w + b}$ in both cases.

---

[5] As defined in Example 3, p. 33.

4

RANDOM VARIABLES

7. Discrete and Continuous Random Variables. Distribution Functions

Given a sample space Ω, by a *random variable* we mean a numerical function $\xi = \xi(\omega)$ whose value depends on the elementary events $\omega \in \Omega$. Let $\mathbf{P}\{x' \leqslant \xi \leqslant x''\}$ be the probability of the event $\{x' \leqslant \xi \leqslant x''\}$, i.e., the probability that ξ takes a value in the interval $x' \leqslant x \leqslant x''$. Then knowledge of $\mathbf{P}\{x' \leqslant \xi \leqslant x''\}$ for all x' and x'' $(x' \leqslant x'')$ is said to specify the *probability distribution* of the random variable ξ.

A random variable $\xi = \xi(\omega)$ is said to be *discrete* (or to have a *discrete distribution*) if ξ takes only a finite or countably infinite number of distinct values x, with corresponding probabilities

$$P_\xi(x) = \mathbf{P}\{\xi = x\},$$

$$\sum_{-\infty}^{\infty} P_\xi(x) = 1,$$

where the summation is over all the values of x taken by ξ. For such random variables,

$$\mathbf{P}\{x' \leqslant \xi \leqslant x''\} = \sum_{x'}^{x''} P_\xi(x), \tag{4.1}$$

where the summation is over the finite or countably infinite number of values of x which ξ can take in the interval $x' \leqslant x \leqslant x''$.

A random variable $\xi = \xi(\omega)$ is said to be *continuous* (or to have a *continuous distribution*) if

$$\mathbf{P}\{x' \leqslant \xi \leqslant x''\} = \int_{x'}^{x''} p_\xi(x)\, dx, \tag{4.2}$$

where $p_\xi(x)$ is a nonnegative integrable function, called the *probability density* of the random variable ξ, with unit integral

$$\int_{-\infty}^{\infty} p_\xi(x)\, dx = 1.$$

Clearly, if ξ is a continuous random variable, then

$$\mathbf{P}\{\xi = x\} = 0$$

for any given value x, while[1]

$$\mathbf{P}\{\xi \in dx\} \sim p_\xi(x)\, dx$$

for every x with a neighborhood in which the probability density $p_\xi(x)$ is continuous. Here $\mathbf{P}\{\xi \in dx\}$ is the probability of the event $\{\xi \in dx\}$, consisting of ξ taking any value in an infinitesimal interval dx centered at the point x.

The function

$$\Phi_\xi(x) = \mathbf{P}\{\xi \leqslant x\}, \qquad -\infty < x < \infty$$

is called the *distribution function* of the random variable ξ. If ξ is a discrete random variable, $\Phi_\xi(x)$ is the step function

$$\Phi_\xi(x) = \sum_{-\infty}^{x} p_\xi(x),$$

taking a finite or countably infinite number of distinct values [the graph of such a function is shown in Figure 4(a)]. If ξ is a continuous random variable, $\Phi_\xi(x)$ is the continuous function

$$\Phi_\xi(x) = \int_{-\infty}^{x} p_\xi(x)\, dx \tag{4.3}$$

[the graph of such a function is shown in Figure 4(b)].[2] Clearly,

$$\mathbf{P}\{x' < \xi \leqslant x''\} = \Phi_\xi(x'') - \Phi_\xi(x') \tag{4.4}$$

for any random variable ξ.

Now consider two random variables ξ_1 and ξ_2, or equivalently, the random point or vector $\xi = (\xi_1, \xi_2)$. First suppose ξ_1 and ξ_2 are discrete. Then ξ_1 and ξ_2 have a *joint probability distribution*, characterized by the probabilities

$$p_{\xi_1,\xi_2}(x_1, x_2) = \mathbf{P}\{\xi_1 = x_1, \xi_2 = x_2\}, \tag{4.5}$$

where x_1 and x_2 range over all possible values of the corresponding random

---

[1] The symbol $\in$ means "belongs to" or "is contained in."

[2] By a well-known theorem on differentiation, $d\Phi_\xi(x)/dx = p_\xi(x)$ almost everywhere. See e.g., E. C. Titchmarsh, *The Theory of Functions*, second edition, Oxford University Press, London (1939), p. 362.

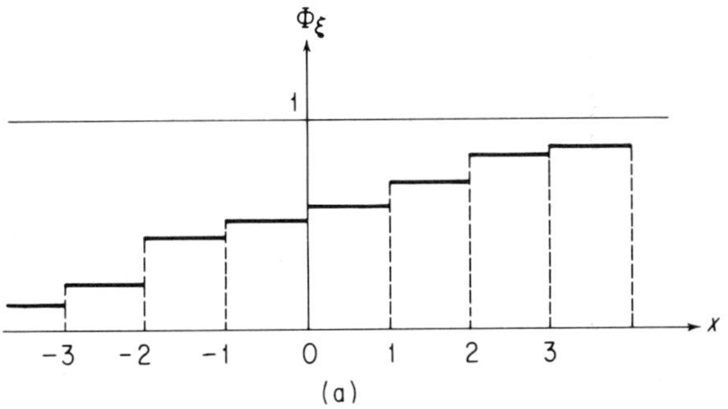

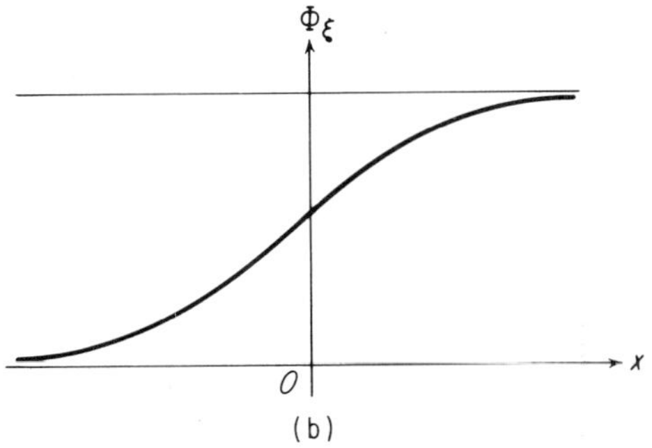

FIGURE 4. (a) A typical distribution function of a discrete random variable taking only the integral values $\ldots, -2, -1, 0, 1, 2, \ldots$ At the points $x = \ldots, -2, -1, 0, 1, 2, \ldots, \Phi_\xi(x)$ has jumps equal to the corresponding probabilities $P_\xi(x)$. (b) A typical distribution function of a continuous random variable. Any continuous monotonic function $\Phi_\xi(x)$ such that $\lim_{x \to -\infty} \Phi_\xi(x) = 0$, $\lim_{x \to +\infty} \Phi_\xi(x) = 1$ can serve as the distribution function of a continuous random variable ξ.[3]

variables ξ_1 and ξ_2. The probability of any event of the type $\{(\xi_1, \xi_2) \in B\}$, i.e., the "probability of the random point $\xi = (\xi_1, \xi_2)$ falling in a given set B," is given by

$$\mathbf{P}\{(\xi_1, \xi_2) \in B\} = \sum_{(x_1, x_2) \in B} p_{\xi_1, \xi_2}(x_1, x_2),$$

where the summation is over all possible values x_1, x_2 of the random variables ξ_1, ξ_2 such that the point (x_1, x_2) lies in B. Next suppose ξ_1 and ξ_2 are continuous. Then by the *joint probability density* of ξ_1 and ξ_2, we mean a

[3] It should be noted that there are random variables which are neither discrete nor continuous but a "mixture of both." There are also continuous distribution functions more general than (4.3).

function $p_{\xi_1,\xi_2}(x_1, x_2)$ of two variables x_1 and x_2 such that the probability of any event of the type $\{(\xi_1, \xi_2) \in B\}$ is given by

$$\mathbf{P}\{(\xi_1, \xi_2) \in B\} = \iint_B p_{\xi_1,\xi_2}(x_1, x_2)\, dx_1\, dx_2 \tag{4.6}$$

(the integral is over B).

Given a family of random variables $\xi_1, \ldots, \xi_n$, suppose the events $\{x_k' \leqslant \xi_k \leqslant x_k''\}$, $k = 1, \ldots, n$ are independent for arbitrary x_k' and x_k'' $(x_k' \leqslant x_k'')$. Then the random variables $\xi_1, \ldots, \xi_n$ are said to be (*statistically*) *independent*. Given an infinite sequence of random variables $\xi_1, \xi_2, \ldots,$ suppose the random variables $\xi_1, \ldots, \xi_n$ are independent for every n, or equivalently that the events $\{x_k' \leqslant \xi_k \leqslant x_k''\}$, $k = 1, 2, \ldots$ are independent for arbitrary x_k' and x_k''. Then $\xi_1, \xi_2, \ldots$ is said to be a *sequence of independent random variables*.

Suppose two random variables ξ_1 and ξ_2 are independent. Then clearly their joint probability distribution (4.5) is such that

$$P_{\xi_1,\xi_2}(x_1, x_2) = P_{\xi_1}(x_1)P_{\xi_2}(x_2) \tag{4.7}$$

if ξ_1 and ξ_2 are discrete, and

$$p_{\xi_1,\xi_2}(x_1, x_2) = p_{\xi_1}(x_1)p_{\xi_2}(x_2) \tag{4.7$'$}$$

if ξ_1 and ξ_2 are continuous. In (4.7′), $p_{\xi_1}(x_1)$ is the probability density of ξ_1 and $p_{\xi_2}(x_2)$ that of ξ_2, while $p_{\xi_1,\xi_2}(x_1, x_2)$ is the joint probability density of ξ_1 and ξ_2 figuring in (4.6).

***Example* 1 (*The uniform distribution*).** Suppose a point ξ is "tossed at random" onto the interval $[a, b]$. This means that the probability of ξ falling in a subinterval $[x', x''] \subset [a, b]$ does not depend on the location of $[x', x'']$. Hence the probability of ξ falling in $[x', x'']$ is proportional to the length $x'' - x'$.[4] More exactly, we have

$$\mathbf{P}\{x' \leqslant \xi \leqslant x''\} = \frac{x'' - x'}{b - a} = \int_{x'}^{x''} \frac{dx}{b - a},$$

since then the probability of ξ falling in $[a, b]$ itself is

$$\mathbf{P}\{a \leqslant \xi \leqslant b\} = \int_a^b \frac{dx}{b - a} = 1,$$

as it must be. Clearly, ξ is a continuous random variable, with probability

---

[4] Let $f(s)$ be the probability of ξ falling in a subinterval of length s. Then clearly $f(s + t) = f(s) + f(t)$. But it can be shown that any function $f(s)$ satisfying this equation is either of the form ks (k a constant) or else unbounded in every interval (see W. Feller, *op. cit.*, p. 459).

density

$$p_\xi(x) = \begin{cases} \dfrac{1}{b-a} & \text{if } a \leqslant x \leqslant b, \\ 0 & \text{if } x < a \text{ or } x > b. \end{cases}$$

Such a random variable is said to have a *uniform distribution.*

***Example* 2.** Suppose two points ξ_1 and ξ_2 are tossed at random and *independently* onto a line segment of length L. What is the probability that the distance between the two points does not exceed l?

Solution. Imagine that ξ_1 falls in an interval $[0, L]$ of the x_1-axis, while ξ_2 falls in an interval $[0, L]$ of the x_2-axis, perpendicular to the x_1-axis as in Figure 5. Then the desired probability is just the probability that a point $\xi = (\xi_1, \xi_2)$ tossed at random onto the square $0 \leqslant x_1, x_2 \leqslant L$ will fall in the

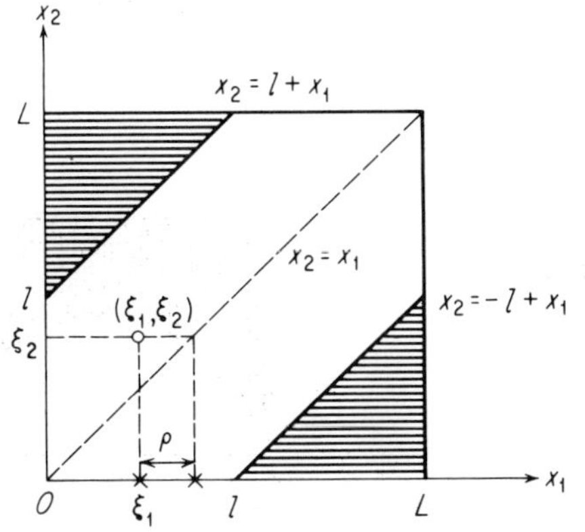

FIGURE 5

region B bounded by the lines $x_2 = l + x_1$ and $x_2 = -l + x_1$ (B is the unshaded region in Figure 5).[5] By hypothesis, the random variables ξ_1 and ξ_2 are independent and are both uniformly distributed in $[0, L]$, i.e., both have probability density

$$p(x) = \frac{1}{L}, \qquad 0 \leqslant x \leqslant L.$$

Hence, by (4.6), the joint probability density of the independent random variables ξ_1 and ξ_2 is just

$$p_{\xi_1,\xi_2}(x_1, x_2) = \frac{1}{L^2}, \qquad 0 \leqslant x_1, x_2 \leqslant L.$$

[5] Note that $|\xi_1 - \xi_2|$ is the horizontal distance between the point (ξ_1, ξ_2) and the line $x_2 = x_1$. This is the distance ρ shown in Figure 5, from which it is apparent that $\rho \leqslant l$ if and only if (ξ_1, ξ_2) lies in B.

Therefore the probability of the random point $\xi = (\xi_1, \xi_2)$ falling in the region B is given by

$$\mathbf{P}\{(\xi_1, \xi_2) \in B\} = \iint_B \frac{dx_1\,dx_2}{L^2} = \frac{2Ll - l^2}{L^2},$$

since $L^2 - 2 \cdot \frac{1}{2}(L - l)^2 = 2Ll - l^2$ is the area of B (the square minus the two shaded triangles).

***Example* 3 (*Buffon's needle problem*).** Suppose a needle is tossed at random onto a plane ruled with parallel lines a distance L apart, where by a "needle" we mean a line segment of length $l < L$. What is the probability of the needle intersecting one of the parallel lines?

Solution. Let ξ_1 be the angle between the needle and the direction of the rulings, and let ξ_2 be the distance between the bottom point of the needle and the nearest line above this point [see Figure 6(a)]. Then the conditions

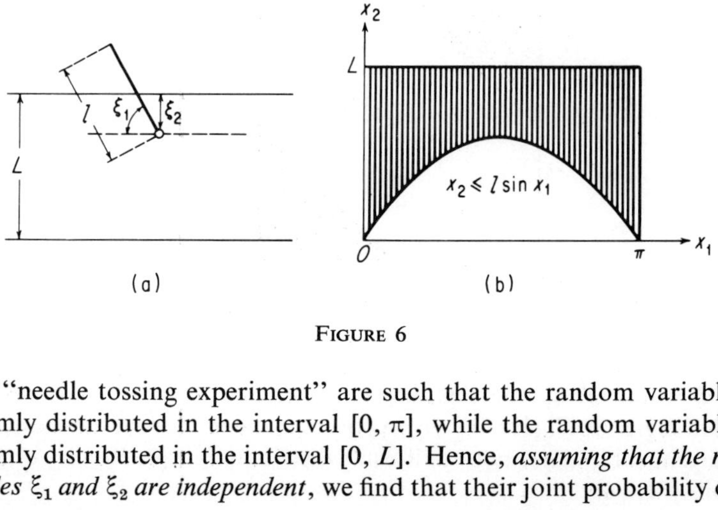

FIGURE 6

of the "needle tossing experiment" are such that the random variable ξ_1 is uniformly distributed in the interval $[0, \pi]$, while the random variable ξ_2 is uniformly distributed in the interval $[0, L]$. Hence, *assuming that the random variables ξ_1 and ξ_2 are independent*, we find that their joint probability density is

$$p_{\xi_1,\xi_2}(x_1, x_2) = \frac{1}{\pi L}, \qquad 0 \leqslant x_1 \leqslant \pi, \qquad 0 \leqslant x_2 \leqslant L.$$

The event consisting of the needle intersecting one of the rulings occurs if and only if

$$\xi_2 \leqslant l \sin \xi_1,$$

i.e., if and only if the corresponding point $\xi = (\xi_1, \xi_2)$ falls in the region B, where B is the part of the rectangle $0 \leqslant x_1 \leqslant \pi$, $0 \leqslant x_2 \leqslant L$ lying between the x_1-axis and the curve $x_2 = \sin x_1$ [B is the unshaded region in Figure

6(b)]. Hence, by the general formula (4.6),

$$\mathbf{P}\{(\xi_1, \xi_2) \in B\} = \iint_B \frac{dx_1\,dx_2}{\pi L} = \frac{2l}{\pi L}, \tag{4.8}$$

where

$$l\int_0^\pi \sin x_1\,dx_1 = 2l$$

is the area of B.

In deducing (4.8), we have assumed that ξ_1 and ξ_2 are independent random variables. This assumption can be tested experimentally. In fact, according to (4.8), if the needle is repeatedly tossed onto the ruled plane, then the frequency of the event A, consisting of the needle intersecting one of the rulings, must be approximately $2l/\pi L$. Suppose the needle is tossed n times, and let $n(A)$ be the number of times A occurs, so that $n(A)/n$ is the relative frequency of the event A. Then

$$\frac{n(A)}{n} \sim \frac{2l}{\pi L}$$

for large n, as discussed on p. 3. Hence

$$\frac{2l}{L}\frac{n}{n(A)}$$

should be a good approximation to $\pi = 3.14\ldots$ for large n. This actually turns out to be the case.[6]

***Example* 4.** Given two independent random variables ξ_1 and ξ_2, with probability densities $p_{\xi_1}(x_1)$ and $p_{\xi_2}(x_2)$, find the probability density of the random variable

$$\eta = \xi_1 + \xi_2.$$

Solution. By (4.7′), the joint probability distribution of ξ_1 and ξ_2 equals $p_{\xi_1}(x_1)p_{\xi_2}(x_2)$, and hence, by (4.6),

$$P\{y' \leqslant \eta \leqslant y''\} = \iint_{y' \leqslant x_1 + x_2 \leqslant y''} p_{\xi_1}(x_1)p_{\xi_2}(x_2)\,dx_1\,dx_2$$

$$= \int_{y'}^{y''} \left[\int_{-\infty}^{\infty} p_{\xi_1}(y - x)p_{\xi_2}(x)\,dx\right] dy.$$

Therefore the probability density of the random variable η is given by the expression

$$p_\eta(y) = \int_{-\infty}^{\infty} p_{\xi_1}(y - x)p_{\xi_2}(x)\,dx,$$

called the *composition* or *convolution* of the functions p_{ξ_1} and p_{ξ_2}.

[6] See J. V. Uspensky, *Introduction to Mathematical Probability*, McGraw-Hill Book Co., Inc., New York (1937), p. 113.

For example, suppose ξ_1 and ξ_2 are both uniformly distributed in the interval [0, 1], so that they both have the probability density

$$p(x) = \begin{cases} 1 & \text{if } 0 \leqslant x \leqslant 1, \\ 0 & \text{if } x < 0 \text{ or } x > 1. \end{cases}$$

Then

$$p_\eta(y) = \begin{cases} \int_0^y dx = y & \text{if } 0 \leqslant y \leqslant 1, \\ \int_{y-1}^1 dx = 2 - y & \text{if } 1 \leqslant y \leqslant 2, \\ 0 & \text{if } y < 0 \text{ or } y > 2. \end{cases}$$

The graph of the density $p_\eta(y)$ is triangular in shape, as shown in Figure 7.

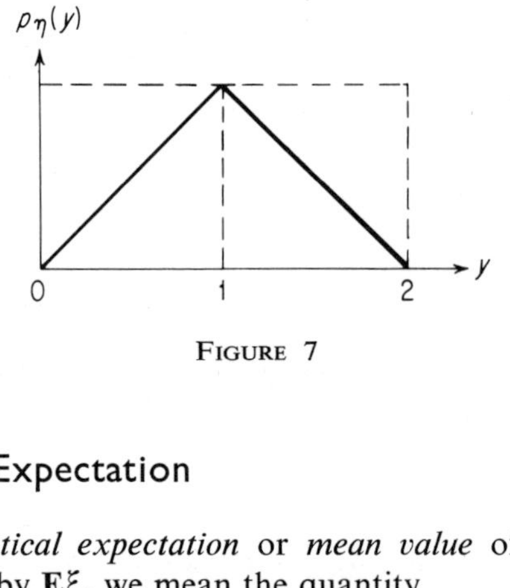

FIGURE 7

8. Mathematical Expectation

By the *mathematical expectation* or *mean value* of a discrete random variable ξ, denoted by $\mathbf{E}\xi$, we mean the quantity

$$\mathbf{E}\xi = \sum_{-\infty}^{\infty} x P_\xi(x), \tag{4.9}$$

provided that the series converges absolutely.[7] Here the summation has the same meaning as on p. 37, and, as usual, $\mathbf{P}_\xi(x) = \mathbf{P}\{\xi = x\}$. Given a discrete random variable ξ, consider the new random variable $\eta = \varphi(\xi)$, where $\varphi(x)$ is some function of x. Then the mathematical expectation of η is given in terms of the probability distribution of ξ by the formula

$$\mathbf{E}\eta = \mathbf{E}\varphi(\xi) = \sum_{-\infty}^{\infty} \varphi(x) P_\xi(x). \tag{4.10}$$

---

[7] I.e., provided that $\sum_{-\infty}^{\infty} |x|\, P_\xi(x) < \infty$.

In fact, η is a discrete random variable taking only the values $y = \varphi(x)$, where x ranges over all possible values of the random variable ξ. Therefore[8]

$$\mathbf{P}\{\eta = y\} = \sum_{x:\varphi(x)=y} P_\xi(x),$$

where the summation is over all x such that $\varphi(x) = y$, and hence

$$\mathbf{E}\eta = \sum_{-\infty}^{\infty} y\mathbf{P}\{\eta = y\} = \sum_{-\infty}^{\infty} \sum_{x:\varphi(x)=y} P_\xi(x) = \sum_{-\infty}^{\infty} \varphi(x)P_\xi(x),$$

as asserted.

More generally, let $\varphi(\xi_1, \xi_2)$ be a random variable which is a function of two random variables ξ_1 and ξ_2, with joint probability distribution $P_{\xi_1,\xi_2}(x_1, x_2)$. Then it is easily verified that $\varphi(\xi_1, \xi_2)$ has the mathematical expectation

$$\mathbf{E}\varphi(\xi_1, \xi_2) = \sum_{-\infty}^{\infty} \sum_{-\infty}^{\infty} \varphi(x_1, x_2)P_{\xi_1,\xi_2}(x_1, x_2). \tag{4.11}$$

It is clear from (4.9) that

a) $\mathbf{E}1 = 1$;
b) $\mathbf{E}(c\xi) = c\mathbf{E}\xi$ for an arbitrary constant c;
c)
$$|\mathbf{E}\xi| \leqslant \mathbf{E}\,|\xi|. \tag{4.12}$$

Moreover, it follows from (4.11) that

d) $\mathbf{E}(\xi_1 + \xi_2) = \mathbf{E}\xi_1 + \mathbf{E}\xi_2$ for arbitrary random variables ξ_1 and ξ_2 with mathematical expectations $\mathbf{E}\xi_1$ and $\mathbf{E}\xi_2$;
e) $\xi \geqslant 0$ implies $\mathbf{E}\xi \geqslant 0$, and more generally

$$\xi_1 \leqslant \xi_2 \quad \text{implies} \quad \mathbf{E}\xi_1 \leqslant \mathbf{E}\xi_2; \tag{4.13}$$

f) If ξ_1 and ξ_2 are independent random variables, then

$$\mathbf{E}\xi_1\xi_2 = \mathbf{E}\xi_1\,\mathbf{E}\xi_2. \tag{4.14}$$

For example, to prove (4.14), we write $\varphi(\xi_1, \xi_2) = \xi_1\xi_2$. Then, for independent ξ_1 and ξ_2, (4.11) implies

$$\begin{aligned}\mathbf{E}(\xi_1\xi_2) &= \sum_{-\infty}^{\infty} \sum_{-\infty}^{\infty} x_1x_2P_{\xi_1}(x_1)P_{\xi_2}(x_2)\\ &= \sum_{-\infty}^{\infty} x_1P_{\xi_1}(x_1) \sum_{-\infty}^{\infty} x_2P_{\xi_2}(x_2) = \mathbf{E}\xi_1\mathbf{E}\xi_2.\end{aligned}$$

To define the mathematical expectation of a *continuous* random variable ξ, we first approximate ξ by a sequence of discrete random variables ξ_n,

---

[8] The colon should be read as "such that."

$n = 1, 2, \ldots$ Let ε_n, $n = 1, 2, \ldots$ be a sequence of positive numbers converging to zero. Then for each $n = 1, 2, \ldots$, let

$$\ldots, x_{-2,n}, x_{-1,n}, x_{0,n}, x_{1,n}, x_{2,n}, \ldots \tag{4.15}$$

be an infinite set of distinct points such that[9]

$$\sup_k |x_{k,n} - x_{k-1,n}| = \varepsilon_n, \tag{4.16}$$

and let ξ_n be a discrete random variable such that

$$\xi_n = x_{k,n} \qquad \text{if} \quad x_{k-1,n} < \xi \leqslant x_{k,n}.$$

It follows that

$$|\xi - \xi_n| \leqslant \varepsilon_n,$$

and hence

$$|\xi_m - \xi_n| \leqslant |\xi_m - \xi_n| + |\xi_n - \xi| \leqslant \varepsilon_m + \varepsilon_n \to 0$$

as $m, n \to \infty$. Therefore, by (4.12) and (4.13),

$$|\mathbf{E}\xi_m - \mathbf{E}\xi_n| = |\mathbf{E}(\xi_m - \xi_n)| \leqslant \mathbf{E}\,|\xi_m - \xi_n| \leqslant \varepsilon_m + \varepsilon_n \to 0$$

as $m, n \to \infty$ (provided $\mathbf{E}\xi_n$ exists for all n). But then

$$\lim_{n \to \infty} \mathbf{E}\xi_n$$

exists, by the Cauchy convergence criterion. This limit is called the *mathematical expectation* or *mean value* of the continuous random variable ξ, again denoted by $\mathbf{E}\xi$. Clearly,

$$\mathbf{E}\xi = \lim_{n \to \infty} \sum_{-\infty}^{\infty} x_{k,n} \mathbf{P}\{x_{k-1,n} < \xi \leqslant x_{k,n}\}.$$

Suppose ξ has the probability density $p_\xi(x)$. Then, choosing the points (4.15) to be continuity points of $p_\xi(x)$, we have

$$\sum_{-\infty}^{\infty} x_{k,n} \mathbf{P}\{x_{k-1,n} < \xi \leqslant x_{k,n}\} = \sum_{-\infty}^{\infty} x_{k,n} \int_{x_{k-1,n}}^{x_{k,n}} p_\xi(x)\,dx$$
$$\sim \sum_{-\infty}^{\infty} x_{k,n} p_\xi(x_{k,n})(x_{k,n} - x_{k-1,n}),$$

and hence

$$\mathbf{E}\xi = \int_{-\infty}^{\infty} x p_\xi(x)\,dx \tag{4.17}$$

[compare this with (4.9)]. For continuous random variables of the form

---

[9] The symbol sup denotes the supremum or least upper bound. Therefore the left-hand side of (4.16) is the least upper bound of all the differences $|x_{k,n} - x_{k-1,n}|$, $k = \ldots, -2, -1, 0, 1, 2, \ldots$ Thus (4.16) means that no two of the points (4.15) are more than ε_n apart. Note also that any closed interval of length ε_n contains at least two of the points (4.15).

$\eta = \varphi(\xi)$ and $\eta = \varphi(\xi_1, \xi_2)$, we have

$$\mathbf{E}\varphi(\xi) = \int_{-\infty}^{\infty} \varphi(x) p_\xi(x)\, dx \tag{4.18}$$

and

$$\mathbf{E}\varphi(\xi_1, \xi_2) = \int_{-\infty}^{\infty}\int_{-\infty}^{\infty} \varphi(x_1, x_2) p_{\xi_1,\xi_2}(x_1, x_2)\, dx_1\, dx_2, \tag{4.19}$$

by analogy with (4.10) and (4.11), where $p_{\xi_1,\xi_2}(x_1, x_2)$ is the joint probability density of the random variables ξ_1 and ξ_2. It is easily verified that properties a)–f) of the mathematical expectation continue to hold for continuous random variables (the details are left as an exercise).

Remark. Other synonyms for the mathematical expectation of a random variable ξ, discrete or continuous, are the *expected value* of ξ or the *average* (*value*) of ξ. The mathematical expectation and mean value are often simply called the "expectation" and "mean," respectively.

Example Let ξ be a random variable uniformly distributed in the interval $[a, b]$, i.e., let ξ have the probability density

$$p_\xi(x) = \begin{cases} \dfrac{1}{b-a} & \text{if} \quad a \leqslant x \leqslant b, \\ 0 & \text{if} \quad x < a \quad \text{or} \quad x > b. \end{cases}$$

Then the mathematical expectation of ξ is

$$\mathbf{E}\xi = \int_a^b \frac{x\, dx}{b-a} = \frac{a+b}{2}.$$

A random variable of the form $\eta = \xi_1 + i\xi_2$ involving two real random variables ξ_1 and ξ_2 (the real and imaginary parts of η) is called a *complex random variable*. The mathematical expectation of $\eta = \xi_1 + i\xi_2$ is defined as

$$\mathbf{E}\eta = \mathbf{E}\xi_1 + i\mathbf{E}\xi_2.$$

It is easy to see that formulas (4.10) and (4.18) remain valid for the case where $\varphi(\xi)$ is a complex-valued function of a real random variable ξ, and that (4.11) and (4.19) remain valid for the case where $\varphi(\xi_1, \xi_2)$ is a complex-valued function of two real random variables ξ_1 and ξ_2. In particular, let $\varphi_1(\xi_1)$ and $\varphi_2(\xi_2)$ be complex-valued functions of two *independent* real random variables ξ_1 and ξ_2. Then, choosing $\varphi(\xi_1, \xi_2) = \varphi_1(\xi_1)\varphi_2(\xi_2)$ in (4.11) or (4.19), we deduce the formula

$$\mathbf{E}[\varphi_1(\xi_1)\varphi_2(\xi_2)] = \mathbf{E}\varphi_1(\xi_1)\mathbf{E}\varphi_2(\xi_2), \tag{4.20}$$

which generalizes (4.14).

9. Chebyshev's Inequality. The Variance and Correlation Coefficient

By the *mean square value* of a (real) random variable ξ is meant the quantity $\mathbf{E}\xi^2$, equal to

$$\mathbf{E}\xi^2 = \sum_{-\infty}^{\infty} x^2 P_\xi(x)$$

if ξ is discrete, or

$$\mathbf{E}\xi^2 = \int_{-\infty}^{\infty} x^2 p_\xi(x)\, dx$$

if ξ is continuous.[10] Given any random variable ξ and any number $\varepsilon > 0$, let

$$\xi_1 = \begin{cases} 0 & \text{if } |\xi| < \varepsilon, \\ \varepsilon^2 & \text{if } |\xi| > \varepsilon. \end{cases}$$

Then obviously $\xi_1 \leqslant \xi^2$, and hence, by (4.13),

$$\mathbf{E}\xi_1 \leqslant \mathbf{E}\xi^2,$$

or equivalently

$$\varepsilon^2\mathbf{P}\{|\xi| > \varepsilon\} \leqslant \mathbf{E}\xi^2,$$

since clearly

$$\mathbf{E}\xi_1 = \varepsilon^2\mathbf{P}\{|\xi| > \varepsilon\}.$$

It follows that

$$\mathbf{P}\{|\xi| > \varepsilon\} \leqslant \frac{1}{\varepsilon^2}\mathbf{E}\xi^2, \tag{4.21}$$

a result known as *Chebyshev's inequality*. According to (4.21), if $\mathbf{E}\xi^2/\varepsilon^2 \leqslant \delta$, then $\mathbf{P}\{|\xi| > \varepsilon\} \leqslant \delta$, and hence $\mathbf{P}\{|\xi| \leqslant \varepsilon\} \geqslant 1 - \delta$. Therefore, if δ is small, it is highly probable that $|\xi| \leqslant \varepsilon$. In particular, if $\mathbf{E}\xi^2 = 0$, then $\mathbf{P}\{|\xi| > \varepsilon\} = 0$ for every $\varepsilon > 0$, and hence $\xi = 0$ with probability 1.

By the *variance* or *dispersion* of a random variable ξ, denoted by $\mathbf{D}\xi$, we mean the mean square value $\mathbf{E}(\xi - a)^2$ of the difference $\xi - a$, where $a = \mathbf{E}\xi$ is the mean value of ξ. It follows from

$$\mathbf{E}(\xi - a)^2 = \mathbf{E}\xi^2 - 2a\mathbf{E}\xi + a^2 = \mathbf{E}\xi^2 - 2a^2 + a^2$$

that

$$\mathbf{D}\xi = \mathbf{E}\xi^2 - a^2.$$

Obviously

$$\mathbf{D}1 = 0,$$

and

$$\mathbf{D}(c\xi) = c^2\mathbf{D}\xi$$

for an arbitrary constant c.

---

[10] It is assumed, of course, that $\mathbf{E}\xi^2$ exists. This is not always the case (see e.g., Problem 24, p. 53).

If ξ_1 and ξ_2 are independent random variables, then

$$\mathbf{D}(\xi_1 + \xi_2) = \mathbf{D}\xi_1 + \mathbf{D}\xi_2.$$

In fact, if $a_1 = \mathbf{E}\xi_1$ and $a_2 = \mathbf{E}\xi_2$, then, by (4.14),

$$\mathbf{E}(\xi_1 - a_1)(\xi_2 - a_2) = \mathbf{E}(\xi_1 - a_1)\mathbf{E}(\xi_2 - a_2) = 0, \tag{4.22}$$

and hence

$$\begin{aligned}\mathbf{D}(\xi_1 + \xi_2) &= \mathbf{E}(\xi_1 + \xi_2 - a_1 - a_2)^2 \\ &= \mathbf{E}(\xi_1 - a_1)^2 + 2\mathbf{E}(\xi_1 - a_1)(\xi_2 - a_2) + \mathbf{E}(\xi_2 - a_2)^2 \\ &= \mathbf{E}(\xi_1 - a_1)^2 + \mathbf{E}(\xi_2 - a_2)^2 = \mathbf{D}\xi_1 + \mathbf{D}\xi_2.\end{aligned}$$

Given two random variables ξ_1 and ξ_2, we now consider the problem of finding the linear expression of the form $\hat{c}_1 + \hat{c}_2\xi_2$, involving constants $\hat{c}_1$ and $\hat{c}_2$, such that $\hat{c}_1 + \hat{c}_2\xi_2$ is the best "mean square approximation" to ξ_1, in the sense that

$$\mathbf{E}\,(\xi_1 - \hat{c}_1 - \hat{c}_2\xi_2)^2 = \min_{c_1, c_2} \mathbf{E}\,(\xi_1 - c_1 - c_2\xi_2)^2, \tag{4.23}$$

where the minimum is taken with respect to all c_1 and c_2. To solve this problem, we let

$$\begin{aligned} a_1 &= \mathbf{E}\xi_1, \qquad \sigma_1^2 = \mathbf{D}\xi_1, \\ a_2 &= \mathbf{E}\xi_2, \qquad \sigma_1^2 = \mathbf{D}\xi_2, \end{aligned} \tag{4.24}$$

and introduce the quantity

$$r = \frac{\mathbf{E}(\xi_1 - a_1)(\xi_2 - a_2)}{\sigma_1\sigma_2}, \tag{4.25}$$

called the *correlation coefficient* of the random variables ξ_1 and ξ_2. Going over for convenience to the "normalized" random variables

$$\eta_1 = \frac{\xi_1 - a_1}{\sigma_1}, \qquad \eta_2 = \frac{\xi_2 - a_2}{\sigma_2},$$

we find that

$$\min_{c_1, c_2} \mathbf{E}(\xi_1 - c_1 - c_2\xi_2)^2 = \sigma_1^2 \min_{c_1, c_2} \mathbf{E}(\eta_1 - c_1 - c_2\eta_2)^2 \tag{4.26}$$

(why?). Clearly,

$$\begin{aligned} &\mathbf{E}\eta_1 = \mathbf{E}\eta_2 = 0, \quad \mathbf{D}\eta_1 = \mathbf{E}\eta_1^2 = \mathbf{D}\eta_2 = \mathbf{E}\eta_2^2 = 1, \quad \mathbf{E}\eta_1\eta_2 = r, \\ &\mathbf{E}(\eta_1 - r\eta_2)\eta_2 = \mathbf{E}\eta_1\eta_2 - r\mathbf{E}\eta_2^2 = 0, \\ &\mathbf{E}(\eta_1 - r\eta_2)^2 = \mathbf{E}\eta_1^2 - 2r\mathbf{E}\eta_1\eta_2 + r^2\mathbf{E}\eta_2^2 = 1 - r^2, \end{aligned}$$

and hence

$$\begin{aligned}\mathbf{E}(\eta_1 - c_1 - c_2\eta_2)^2 &= \mathbf{E}[(\eta_1 - r\eta_2) - c_1 + (r - c_2)\eta_2]^2 \\ &= (1 - r^2) + c_1^2 + (r - c_2)^2\end{aligned}$$

for arbitrary c_1 and c_2. It follows that the minimum of $\mathbf{E}(\eta_1 - c_1 - c_2\eta_2)^2$ is achieved for $c_1 = 0$, $c_2 = r$, and is equal to

$$\min_{c_1, c_2} \mathbf{E}(\eta_1 - c_1 - c_2\eta_2)^2 = 1 - r^2. \tag{4.27}$$

But

$$\eta_1 - r\eta_2 = \frac{1}{\sigma_1}\left[\xi_1 - a_1 - r\frac{\sigma_1}{\sigma_2}(\xi_2 - a_2)\right]$$

in terms of the original random variables ξ_1 and ξ_2. Therefore

$$\hat{c}_1 + \hat{c}_2\xi_2 = a_1 + r\frac{\sigma_1}{\sigma_2}(\xi_2 - a_2)$$

is the minimizing linear expression figuring in (4.23), where a_1, a_2, σ_1^2, σ_2^2 and r are defined by (4.24) and (4.25).

If ξ_1 and ξ_2 are *independent*, then, by (4.22),

$$r = \frac{\mathbf{E}(\xi_1 - a_1)(\xi_2 - a_2)}{\sigma_1\sigma_2} = \frac{\mathbf{E}(\xi_1 - a_1)\mathbf{E}(\xi_2 - a_2)}{\sigma_1\sigma_2} = 0.$$

It is clear from (4.27) that r lies in the interval $-1 \leqslant r \leqslant 1$. Moreover, if $r = \pm 1$, then the random variable ξ_1 is simply a linear expression of the form

$$\xi_1 = \hat{c}_1 + \hat{c}_2\xi_2.$$

In fact, if $r = \pm 1$, then, by (4.26) and (4.27), the mean square value of $\xi_1 - \hat{c}_1 - \hat{c}_2\xi_2$ is just

$$\mathbf{E}(\xi_1 - \hat{c}_1 - \hat{c}_2\xi_2)^2 = \sigma_1^2(1 - r^2) = 0,$$

and hence $\xi_1 - \hat{c}_1 - \hat{c}_2\xi_2 = 0$ with probability 1 (why?).

The above considerations seem to suggest the use of r as a measure of the extent to which the random variables ξ_1 and ξ_2 are dependent. However, although suitable in some situations (see Problem 15, p. 67), this use of r is not justified in general (see Problem 19, p. 53).[11]

PROBLEMS

1. A motorist encounters four consecutive traffic lights, each equally likely to be red or green. Let ξ be the number of green lights passed by the motorist before being stopped by a red light. What is the probability distribution of ξ?

2. Give an example of two distinct random variables with the same distribution function.

3. Find the distribution function of the uniformly distributed random variable ξ considered in Example 1, p. 40.

---

[11] See also W. Feller, *op. cit.*, p. 236.

4. A random variable ξ has probability density

$$p_\xi(x) = \frac{a}{x^2 + 1} \qquad (-\infty < x < \infty).$$

Find

a) The constant a; b) The distribution function of ξ;
c) The probability $\mathbf{P}\{-1 \leqslant \xi \leqslant 1\}$.

Ans. a) $\dfrac{1}{\pi}$; b) $\dfrac{1}{2} + \dfrac{1}{\pi}$ arc tan x; c) $\dfrac{1}{2}$.

5. A random variable ξ has probability density

$$p_\xi(x) = \begin{cases} ax^2e^{-kx} & \text{if } 0 \leqslant x < \infty, \\ 0 & \text{otherwise} \end{cases}$$

($k > 0$). Find
a) The constant a; b) The distribution function of ξ;
c) The probability $\mathbf{P}\{0 \leqslant \xi \leqslant 1/k\}$.

6. A random variable ξ has distribution function

$$\Phi_\xi(x) = a + b \text{ arc tan } \frac{x}{2} \qquad (-\infty < x < \infty).$$

Find
a) The constants a and b; b) The probability density of ξ.

7. Two nonoverlapping circular disks of radius r are painted on a circular table of radius R. A point is then "tossed at random" onto the table. What is the probability of the point falling in one of the disks?

Ans. $2(r/R)^2$.

8. What is the probability that two randomly chosen numbers between 0 and 1 will have a sum no greater than 1 and a product no greater than $\frac{2}{9}$?

Ans. $\dfrac{1}{3} + \dfrac{2}{9}\displaystyle\int_{1/3}^{2/3} \frac{dx}{x} = \dfrac{1}{3} + \dfrac{2}{9} \ln 2 \approx 0.49.$

9. Given two independent random variables ξ_1 and ξ_2, with probability densities

$$p_{\xi_1}(x) = \begin{cases} \frac{1}{2}e^{-x/2} & \text{if } x \geqslant 0, \\ 0 & \text{if } x < 0, \end{cases} \qquad p_{\xi_2}(x) = \begin{cases} \frac{1}{3}e^{-x/3} & \text{if } x \geqslant 0, \\ 0 & \text{if } x < 0, \end{cases}$$

find the probability density of the random variable $\eta = \xi_1 + \xi_2$.

Ans. $p_\eta(x) = \begin{cases} e^{-x/3}(1 - e^{-x/6}) & \text{if } x \geqslant 0, \\ 0 & \text{if } x < 0. \end{cases}$

10. Given three independent random variables ξ_1, ξ_2 and ξ_3, each uniformly distributed in the interval [0, 1], find the probability density of the random variable $\xi_1 + \xi_2 + \xi_3$.

Hint. The probability density of $\xi_1 + \xi_2$ (say) was found in Example 4, p. 43.

11. A random variable ξ takes the values $1, 2, \ldots, n, \ldots$ with probabilities

$$\frac{1}{3}, \frac{1}{3^2}, \ldots, \frac{1}{3^n}, \ldots$$

Find $\mathbf{E}\xi$.

12. Balls are drawn from an urn containing w white balls and b black balls until a white ball appears. Find the mean value m and variance σ^2 of the number of black balls drawn, assuming that each ball is replaced after being drawn.

Ans. $m = \dfrac{b}{w}, \quad \sigma^2 = \dfrac{b(w+b)}{w^2}.$

13. Find the mean and variance of the random variable ξ with probability density

$$p_\xi(x) = \tfrac{1}{2}e^{-|x|} \qquad (-\infty < x < \infty).$$

Ans. $\mathbf{E}\xi = 0, \mathbf{D}\xi = 2.$

14. Find the mean and variance of the random variable ξ with probability density

$$p_\xi(x) = \begin{cases} \dfrac{1}{2b} & \text{if } |x - a| \leqslant b, \\ 0 & \text{otherwise.} \end{cases}$$

Ans. $\mathbf{E}\xi = a, \mathbf{D}\xi = b^2/3.$

15. The distribution function of a random variable ξ is

$$\Phi_\xi(x) = \begin{cases} 0 & \text{if } x \leqslant -1, \\ a + b \arcsin x & \text{if } -1 \leqslant x \leqslant 1, \\ 1 & \text{if } x \geqslant 1. \end{cases}$$

Find $\mathbf{E}\xi$ and $\mathbf{D}\xi$.

Hint. First determine a and b.

Ans. $\mathbf{E}\xi = 0, \mathbf{D}\xi = \tfrac{1}{2}.$

16. Let ξ be the number of spots obtained in throwing an unbiased die. Find the mean and variance of ξ.

Ans. $\mathbf{E}\xi = \tfrac{7}{2}, \mathbf{D}\xi = \tfrac{35}{12}.$

17. In the preceding problem, what is the probability P of ξ deviating from $\mathbf{E}\xi$ by more than $\tfrac{5}{2}$? Show that Chebyshev's inequality gives only a very crude estimate of P.

18. Prove that if ξ is a random variable such that $\mathbf{E}e^{a\xi}$ exists, where a is a positive constant, then

$$\mathbf{P}\{\xi > \varepsilon\} \leqslant \frac{\mathbf{E}e^{a\xi}}{e^{a\varepsilon}}$$

Hint. Apply Chebyshev's inequality to the random variable $\eta = e^{a\xi/2}$.

19. Let ξ be a random variable taking each of the values -2, -1, 1 and 2 with probability $\frac{1}{4}$, and let $\eta = \xi^2$. Prove that ξ and η (although obviously dependent) have correlation coefficient 0.

20. Find the means and variances of two random variables ξ_1 and ξ_2 with joint probability density

$$p_{\xi_1,\xi_2}(x_1, x_2) = \begin{cases} \sin x_1 \sin x_2 & \text{if } 0 \leqslant x_1 \leqslant \dfrac{\pi}{2}, \quad 0 \leqslant x_2 \leqslant \dfrac{\pi}{2} \\ 0 & \text{otherwise.} \end{cases}$$

What is the correlation coefficient of ξ_1 and ξ_2?

21. Find the correlation coefficient r of two random variables ξ_1 and ξ_2 with joint probability density

$$p_{\xi_1,\xi_2}(x_1, x_2) = \begin{cases} \dfrac{1}{2} \sin (x_1 + x_2) & \text{if } 0 \leqslant x_1 \leqslant \dfrac{\pi}{2}, \quad 0 \leqslant x_2 \leqslant \dfrac{\pi}{2}, \\ 0 & \text{otherwise.} \end{cases}$$

Ans. $$r = \frac{\dfrac{\pi}{2} - 1 - \dfrac{\pi^2}{16}}{\dfrac{\pi}{2} - 2 + \dfrac{\pi^2}{16}} \approx -\frac{1}{4}.$$

22. Given a random variable ξ, let $\varphi(t)$ be a nondecreasing positive function such that $\mathbf{E}\varphi(\xi)$ exists. Prove that

$$\mathbf{P}\{\xi > t\} \leqslant \frac{m}{\varphi(t)}. \qquad (4.28)$$

23. Deduce Chebyshev's inequality as a special case of (4.28).

24. Let ξ be a random variable with probability density

$$p_\xi(x) = \frac{1}{\pi(1 + x^2)} \qquad (-\infty < x < \infty).$$

Show that $\mathbf{E}\xi$ and $\mathbf{D}\xi$ fail to exist.

5

THREE IMPORTANT PROBABILITY DISTRIBUTIONS

10. Bernoulli Trials. The Binomial and Poisson Distributions

By *Bernoulli trials* we mean identical independent experiments in each of which an event A, say, may occur with probability

$$p = \mathbf{P}(A)$$

$(p \neq 0)$ or fail to occur with probability

$$q = 1 - p.$$

Occurrence of the event A is called a "success," and nonoccurrence of A (i.e., occurrence of the complementary event $\bar{A}$) is called a "failure."

In the case of n consecutive Bernoulli trials, each elementary event ω can be described by a sequence like

$$\underbrace{1011\ldots0001}_{n \text{ times}}$$

consisting of n digits, each a 0 or a 1, where success at the ith trial is denoted by a 1 in the ith place and failure at the ith trial by a 0 in the ith place. Because of the independence of the trials, the probability of an elementary event ω in which there are precisely k successes and $n - k$ failures is just

$$\mathbf{P}(\omega) = p^k q^{n-k}.$$

Clearly, the various elementary events are equiprobable only if $p = q$.

Now consider the random variable ξ equal to the total number of successes in n Bernoulli trials, i.e., $\xi(\omega) = k$ if precisely k successes occur in the elementary event ω. The number of distinct elementary events with the same total number of successes k is just the number of distinct sequences consisting of k ones and $n - k$ zeros. But the number of such sequences is just the binomial coefficient

$$C_k^n = \binom{n}{k} = \frac{n!}{k!\,(n-k)!}, \tag{5.1}$$

equal to the number of combinations of n things taken k at a time (recall Theorem 1.3, p. 7). These C_k^n elementary events all have the same probability

$$\mathbf{P}(\omega) = p^k q^{n-k},$$

and hence the event $\{\xi = k\}$ has probability

$$\mathbf{P}\{\xi = k\} = C_k^n p^k q^{n-k}.$$

Thus the probability distribution of the random variable ξ is given by

$$P_\xi(k) = C_k^n p^k q^{n-k}, \qquad k = 0, 1, \ldots, n, \tag{5.2}$$

and is known as the *binomial distribution*. The binomial distribution is specified by two parameters, the probability p of a single success and the number of trials n.

It should be noted that the random variable ξ is the sum

$$\xi = \xi_1 + \cdots + \xi_n \tag{5.3}$$

of n independent random variables $\xi_1, \ldots, \xi_n$, where $\xi_k = 1$ if "success" occurs at the kth trial and $\xi_k = 0$ if "failure" occurs at the kth trial. We have

$$\mathbf{E}\xi_k = p, \qquad \mathbf{D}\xi_k = \mathbf{E}\xi_k^2 - (\mathbf{E}\xi_k)^2 = p - p^2 = p(1-p) = pq.$$

Therefore

$$\mathbf{E}\xi = np, \qquad \mathbf{D}\xi = npq. \tag{5.4}$$

Suppose the number of trials is large while the probability of success p is relatively small, so that each success is a rather rare event while the average number of successes np is appreciable. Then it is a good approximation to write

$$P_\xi(k) \sim \frac{a^k}{k!} e^{-a}, \qquad k = 0, 1, 2, \ldots, \tag{5.5}$$

where $a = np$ is the average number of successes and $e = 2.718\ldots$ is the base of the natural logarithms. In fact, we know from calculus that

$$\lim_{n \to \infty} \left(1 - \frac{a}{n}\right)^n = e^{-a}.$$

But $p = a/n$, and hence (5.2) gives

$$P_\xi(0) = q^n = \left(1 - \frac{a}{n}\right)^n \sim e^{-a}.$$

Moreover, it is easily found from (5.1) and (5.2) that

$$\frac{P_\xi(k)}{P_\xi(k-1)} = \frac{np - (k-1)p}{kq} \sim \frac{a}{k}$$

as $n \to \infty$. Therefore

$$P_\xi(1) \sim \frac{a}{1} P_\xi(0) \sim \frac{a}{1} e^{-a},$$

$$P_\xi(2) \sim \frac{a}{2} P_\xi(1) \sim \frac{a^2}{1 \cdot 2} e^{-a},$$

$$\cdots\cdots\cdots\cdots$$

$$P_\xi(k) \sim \frac{a}{k} P_\xi(k-1) \sim \frac{a^k}{k!} e^{-a},$$

which proves the approximate formula (5.5).

A random variable ξ taking only the integral values $0, 1, 2, \ldots$ is said to have a *Poisson distribution* if

$$P_\xi(k) = \frac{a^k}{k!} e^{-a}, \qquad k = 0, 1, 2, \ldots \tag{5.6}$$

The distribution (5.6) is specified by a single positive parameter a, equal to the mean value of ξ:

$$a = \mathbf{E}\xi = \sum_{k=0}^{\infty} kP_\xi(k).$$

In fact, it follows from the expansion

$$e^x = \sum_{k=0}^{\infty} \frac{x^k}{k!},$$

valid for all x, that

$$\mathbf{E}\xi = \sum_{k=0}^{\infty} kP_\xi(k) = \sum_{k=0}^{\infty} k \frac{a^k}{k!} e^{-a} = ae^{-a} \sum_{k=1}^{\infty} \frac{a^{k-1}}{(k-1)!} = ae^{-a}e^{a} = a.$$

Remark. Thus the approximate formula (5.5) shows that the total number of successes in n Bernoulli trials has an approximately Poisson distribution with parameter $a = np$, if n is large and the probability of success p is small.

***Example 1* (*The lottery ticket problem*).** How many lottery tickets must be bought to make the probability of winning at least P?

Solution. Let N be the total number of lottery tickets and M the total number of winning tickets. Then M/N is the probability that a bought ticket is one of the winning tickets. The purchase of each ticket can be regarded as a separate trial with probability of "success" $p = M/N$ in a series of n independent trials, where n is the number of tickets bought. If p is relatively small, as is usually the case, and the given probability P is relatively large, then it is clear that a rather large number of tickets must be bought to make the probability of buying at least one winning ticket no smaller than P. Hence the number of winning tickets among those purchased is a random variable with an approximately Poisson distribution, i.e., the probability that there are precisely k winning tickets among the n purchased tickets is

$$P(k) \approx \frac{a^k}{k!} e^{-a},$$

where

$$a = n\frac{M}{N}.$$

The probability that at least one of the tickets is a winning ticket is just

$$1 - P(0) = 1 - e^{-a}.$$

Hence n must be at least as large as the smallest positive integer satisfying the inequality

$$e^{-a} = e^{-nM/N} \leqslant 1 - P.$$

***Example 2* (*The raisin bun problem*).** Suppose N raisin buns of equal size are baked from a batch of dough into which n raisins have been carefully mixed. Then clearly the number of raisins will vary from bun to bun, although the average number of raisins per bun is just $a = n/N$. What is the probability that any given bun will contain at least one raisin?

Solution. It is natural to assume that the volume of the raisins is much less than that of the dough, so that the raisins move around freely and virtually independently during the mixing, and hence whether or not a given raisin ends up in a given bun does not depend on what happens to the other raisins. Clearly, the raisins will be approximately uniformly distributed throughout the dough after careful mixing, i.e., every raisin has the same probability

$$p = \frac{1}{N}$$

of ending up in a given bun.[1] Imagine the raisins numbered from 1 to n,

[1] If v is the volume of the raisins and V that of the dough, then $p = v/V$.

and select a bun at random. Then we can interpret the problem in term as of series of n Bernoulli trials, where "success" at the kth trial means that the kth raisin ends up in the given bun. Suppose both the number of rolls N and the number of raisins n are large, so that in particular $p = 1/N$ is small. Then the number of successes in the n trials, equal to the number of raisins in the given bun, has an approximately Poisson distribution, i.e., the probability $P(k)$ of exactly k raisins appearing in the bun is given by

$$P(k) \approx \frac{a^k}{k!} e^{-a},$$

where

$$a = np = \frac{n}{N}.$$

Hence the probability P of at least one raisin appearing in the bun is

$$P = 1 - P(0) = 1 - e^{-a}.$$

Example 3 (Radioactive decay). It is observed experimentally that radium gradually decays into radon by emitting alpha particles (helium nuclei). The interatomic distances are large enough to justify the assumption that (the nucleus of) each radium atom disintegrates independently of all the others. Moreover, each of the n_0 radium atoms initially present clearly has the same small probability $p(t)$ of disintegrating during an interval of t seconds.[2] Suppose the disintegration of each radium atom is interpreted as a "success." Then the random variable $\xi(t)$, equal to the number of alpha particles emitted in t seconds, equals the number of successes in a series of n_0 Bernoulli trials with probability of success $p(t)$. The values of n_0 and $p(t)$ are such that the distribution of $\xi(t)$ is very accurately a Poisson distribution, i.e., the probability of exactly k alpha particles being emitted is given by

$$\mathbf{P}\{\xi(t) = k\} = \frac{a^k}{k!} e^{-a}, \qquad k = 0, 1, 2, \ldots, \tag{5.7}$$

where

$$a = \mathbf{E}\xi(t) = n_0 p(t)$$

is the average number of alpha particles emitted in t seconds.

Here we have used a model involving Bernoulli trials as a tool for showing that the random variable $\xi(t)$ has a Poisson distribution. Another physical situation leading to a Poisson distribution is considered in Example 4, p. 73.

[2] A gram of radium ($n_0 \approx 10^{22}$) emits about 10^{10} alpha particles per second. Hence $p(1) \approx 10^{10}/10^{22} = 10^{-12}$.

11. The De Moivre-Laplace Theorem. The Normal Distribution

Next we prove the following basic "limit theorem":

THEOREM 5.1 (***De Moivre-Laplace theorem***). *Given n independent identically distributed random variables* $\xi_1, \ldots, \xi_n$, *each taking the value* 1 *with probability p and the value* 0 *with probability* $q = 1 - p$, *let*

$$S_n = \sum_{k=1}^{n} \xi_k, \qquad S_n^* = \frac{S_n - \mathbf{E}S_n}{\sqrt{\mathbf{D}S_n}}.$$

Then

$$\lim_{n\to\infty} \mathbf{P}\{x' \leqslant S_n^* \leqslant x''\} = \frac{1}{\sqrt{2\pi}} \int_{x'}^{x''} e^{-x^2/2}\, dx. \tag{5.8}$$

Proof. S_n is the random variable denoted by ξ in (5.3) and (5.4), i.e., S_n is the number of successes in n Bernoulli trials, with mean and variance

$$\mathbf{E}S_n = np, \qquad \mathbf{D}S_n = npq.$$

Hence the "normalized sum" S_n^* is a random variable taking the values

$$x = \frac{k - np}{\sqrt{npq}}, \qquad k = 0, 1, \ldots, n$$

with probabilities

$$\mathbf{P}\{S_n^* = x\} = P_n(k) = \frac{n!}{k!\,(n-k)!}\, p^k q^{n-k}, \qquad k = 0, 1, \ldots, n.$$

These values divide the interval

$$\left[\frac{-np}{\sqrt{npq}}, \frac{nq}{\sqrt{npq}}\right]$$

into n equal subintervals of length

$$\Delta x = \frac{1}{\sqrt{npq}}.$$

Clearly, as $n \to \infty$,

$$k = np + \sqrt{npq}\, x \to \infty, \qquad n - k = nq - \sqrt{npq}\, x \to \infty,$$

where the convergence is uniform in every finite interval $x' \leqslant x \leqslant x''$. Using Stirling's formula (see p. 10), we find that

$$P_n(k) \sim \frac{\sqrt{2\pi n}\, n^n e^{-n}}{\sqrt{2\pi k}\, k^k e^{-k} \sqrt{2\pi(n-k)}\, (n-k)^{n-k} e^{-(n-k)}}\, p^k q^{n-k}$$

$$= \frac{1}{\sqrt{2\pi}} \sqrt{\frac{n}{k(n-k)}} \left(\frac{np}{k}\right)^k \left(\frac{nq}{n-k}\right)^{n-k}.$$

Moreover,

$$\frac{k}{np} = 1 + \sqrt{\frac{q}{np}}\,x, \qquad \frac{n-k}{nq} = 1 - \sqrt{\frac{p}{nq}}\,x.$$

Therefore, using the expansion

$$\ln(1+\alpha_n) \sim \alpha_n - \frac{\alpha_n^2}{2}$$

(as $\alpha_n \to 0$), we have

$$\ln\left(\frac{k}{np}\right)^{-k} = -k\ln\left(1+\sqrt{\frac{q}{np}}\,x\right)$$

$$\sim -(np+\sqrt{npq}x)\left(\sqrt{\frac{q}{np}}\,x - \frac{1}{2}\frac{q}{np}\,x^2\right),$$

$$\ln\left(\frac{n-k}{nq}\right)^{-(n-k)} = -(n-k)\ln\left(1-\sqrt{\frac{p}{nq}}\,x\right)$$

$$\sim -(nq-\sqrt{npq}\,x)\left(-\sqrt{\frac{p}{nq}}\,x - \frac{1}{2}\frac{p}{nq}\,x^2\right).$$

Adding these expressions, we find that

$$\lim_{n\to\infty}\ln\left(\frac{np}{k}\right)^k\left(\frac{nq}{n-k}\right)^{n-k} = -\frac{x^2}{2},$$

and hence

$$\lim_{n\to\infty}\left(\frac{np}{k}\right)^k\left(\frac{nq}{n-k}\right)^{n-k} = e^{-x^2/2}$$

uniformly in every finite interval $x' \leqslant x \leqslant x''$. Since

$$\sqrt{\frac{n}{n(n-k)}} \sim \sqrt{\frac{k}{np\cdot nq}} = \frac{1}{\sqrt{npq}},$$

it follows that

$$\lim_{n\to\infty}\mathbf{P}\,\{S_n^* = x\} = \frac{1}{\sqrt{2\pi}}\,e^{-x^2/2}\,\Delta x, \qquad \Delta x = \frac{1}{\sqrt{npq}}.$$

Therefore

$$\lim_{n\to\infty}\mathbf{P}\,\{x' \leqslant S_n^* \leqslant x''\} = \lim_{n\to\infty}\sum_{x'\leqslant x\leqslant x''}\mathbf{P}\,\{S_n^* = x\}$$

$$= \lim_{n\to\infty}\sum_{x'\leqslant x\leqslant x''}\frac{1}{\sqrt{2\pi}}\,e^{-x^2/2}\,\Delta x, \qquad (5.9)$$

where the sum is over all values of x in the interval $x' \leqslant x \leqslant x''$. But clearly

$$\lim_{n\to\infty} \sum_{x'\leqslant x\leqslant x''} \frac{1}{\sqrt{2\pi}} e^{-x^2/2} \Delta x = \frac{1}{\sqrt{2\pi}} \int_{x'}^{x''} e^{-x^2/2}\, dx \tag{5.10}$$

(why?). Comparing (5.9) and (5.10), we finally get the desired limiting formula (5.8). ▮

According to Theorem 5.1, the limiting distribution of the random variable S_n^* is the distribution with probability density

$$p(x) = \frac{1}{\sqrt{2\pi}} e^{-x^2/2}. \tag{5.11}$$

Such a distribution is called a *normal* (or *Gaussian*) *distribution*. The density $p(x)$ is the "bell-shaped" curve shown in Figure 8(a). The corresponding distribution function is

$$\Phi(x) = \frac{1}{\sqrt{2\pi}} \int_{-\infty}^{x} e^{-u^2/2}\, du, \tag{5.12}$$

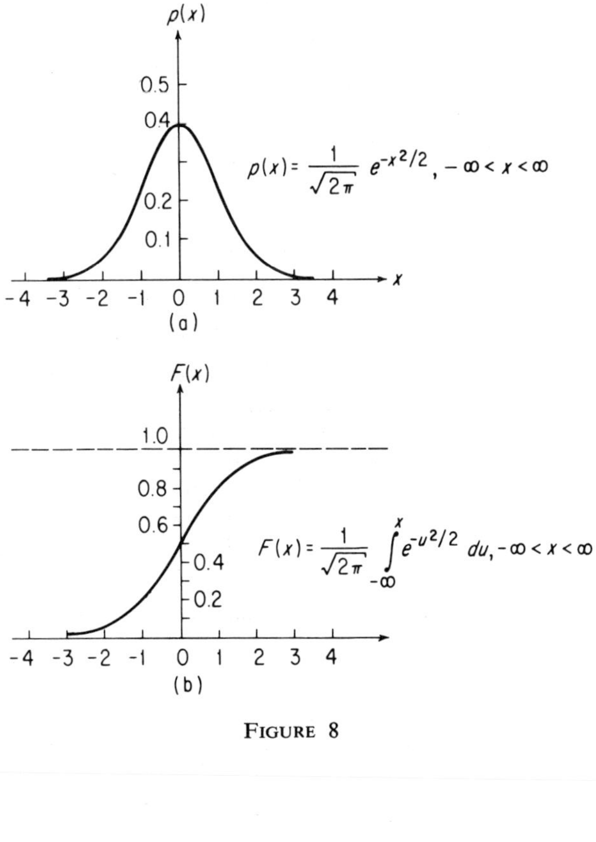

FIGURE 8

Table 2. Values of the normal distribution function $\Phi(x)$ given by formula (5.12).

x	$\Phi(x)$
0.0	0.5000
0.1	0.5398
0.2	0.5793
0.3	0.6179
0.4	0.6554
0.5	0.6915
0.6	0.7257
0.7	0.7580
0.8	0.7881
0.9	0.8159
1.0	0.8413
1.1	0.8643
1.2	0.8849
1.3	0.9032
1.4	0.9192
1.5	0.9332
1.6	0.9452
1.7	0.9554
1.8	0.9641
1.9	0.9713
2.0	0.9773
2.1	0.9821
2.2	0.9861
2.3	0.9893
2.4	0.9918
2.5	0.9938
2.6	0.9953
2.7	0.9965
2.8	0.9974
2.9	0.9981
3.0	0.9986

and is shown in Figure 8(b). Since $p(x)$ is even, it is clear that

$$\Phi(-x) = 1 - \Phi(x).$$

Representative values of $\Phi(x)$ are given in Table 2.

Let ξ be a *normal* (or *Gaussian*) *random variable*, i.e., a random variable with probability density (5.11). Then

$$\mathbf{E}\xi = \frac{1}{\sqrt{2\pi}} \int_{-\infty}^{\infty} x e^{-x^2/2} \, dx = 0,$$

since the integrand is odd. Moreover,

$$\begin{aligned}\mathbf{D}\xi = \mathbf{E}\xi^2 - (\mathbf{E}\xi)^2 = \mathbf{E}\xi^2 &= \frac{1}{\sqrt{2\pi}} \int_{-\infty}^{\infty} x^2 e^{-x^2/2} \, dx \\ &= \frac{1}{\sqrt{2\pi}} \lim_{N \to \infty} \int_{-N}^{N} x^2 e^{-x^2/2} \, dx.\end{aligned}$$

Integrating by parts, we get

$$\begin{aligned}\mathbf{D}\xi &= \frac{1}{\sqrt{2\pi}} \lim_{N \to \infty} \left\{ -\int_{-N}^{N} x d(e^{-x^2/2}) \right\} \\ &= \frac{1}{\sqrt{2\pi}} \lim_{N \to \infty} \left\{ \left[-x e^{-x^2/2} \right]_{x=-N}^{x=N} + \int_{-N}^{N} e^{-x^2/2} \, dx \right\} \\ &= \frac{1}{\sqrt{2\pi}} \int_{-\infty}^{\infty} e^{-x^2/2} \, dx = 1.\end{aligned}$$

Hence ξ has variance 1. More generally, the random variable with probability density

$$p(x) = \frac{1}{\sqrt{2\pi}\,\sigma} e^{-(x-a)^2/2\sigma^2} \tag{5.13}$$

is also called a normal random variable, and has mean a and variance σ^2 (show this).

***Example* (*Brownian motion*).** Suppose a tiny particle is suspended in a homogeneous liquid. Then the particle undergoes random collisions with the molecules of the liquid, and, as a result, moves about continually in a chaotic fashion. This is the phenomenon of *Brownian motion*. As a model of Brownian motion, we make the following simplifying assumptions, characterizing a "discrete random walk" in one dimension:

1) The particle moves only along the x-axis.
2) The particle moves only at the times $t = n\Delta t$, $n = 0, 1, 2, \ldots$
3) Suppose the particle is at position x at time t. Then, regardless of its previous behavior, the particle moves to either of the two neighboring positions $x + \Delta x$ and $x - \Delta x$ ($\Delta x > 0$) with probability $\frac{1}{2}$.

In other words, at each step the particle undergoes a shift of amount Δx either to the right or to the left, with equal probability.[3]

Now let $\xi(t)$ denote the position of our "Brownian particle" at time t, and suppose the particle is at the point $x = 0$ at time $t = 0$, so that $\xi(0) = 0$. Then after $t = n\Delta t$ seconds, the particle undergoes n displacements of amount Δx, of which S_n, say, are to the right (the positive direction) and $n - S_n$ to the left (the negative direction). As a result, the position of the particle at time $t = n\Delta t$ is just

$$\xi(t) = [S_n \Delta x - (n - S_n) \Delta x] = (2S_n - n) \Delta x. \tag{5.14}$$

Moreover, since $\xi(0) = 0$, we have

$$\xi(t) = [\xi(s) - \xi(0)] + [\xi(t) - \xi(s)]$$

for any s in the interval $0 \leqslant s \leqslant t$ (for the time being, s is an integral multiple of Δx). With our assumptions, it is clear that the increments $\xi(s) - \xi(0)$ and $\xi(t) - \xi(s)$ are independent random variables, and that the probability distribution of $\xi(t) - \xi(s)$ is the same as that of $\xi(t - s) - \xi(0)$. Therefore the variance $\sigma^2(t) = \mathbf{D}\xi(t)$ satisfies the relation

$$\sigma^2(t) = \sigma^2(s) + \sigma^2(t - s), \qquad 0 \leqslant s \leqslant t.$$

It follows that $\sigma^2(t)$ is proportional to t, i.e.,[4]

$$\mathbf{D}\xi(t) = \sigma^2 t, \tag{5.15}$$

where σ^2 is a constant called the *diffusion coefficient*. On the other hand, it is easy to see that after a time t, i.e., after $n = t/\Delta t$ steps, the variance of the displacement must be

$$\mathbf{D}\xi(t) = \frac{t}{\Delta t}(\Delta x)^2. \tag{5.16}$$

Comparing (5.15) and (5.16), we obtain

$$\sigma^2 = \frac{(\Delta x)^2}{\Delta t}. \tag{5.17}$$

The displacements of the particle are independent of one another and can be regarded as Bernoulli trials with probability of "success" $p = \frac{1}{2}$, "success" being interpreted as a displacement in the positive direction. In this sense, the number of displacements S_n in the positive direction is just the number of

---

[3] We will eventually pass to the limit $\Delta t \to 0$, $\Delta x \to 0$, thereby getting the "continuous random walk" characteristic of the actual physical process of Brownian motion.

[4] See footnote 4, p. 40.

successes in n Bernoulli trials. Moreover, the relation between the particle's position at time t and the normalized random variable

$$S_n^* = \frac{S_n - np}{\sqrt{npq}} = \frac{1}{\sqrt{n}}(2S_n - n)$$

is given by

$$\xi(t) = S_n^* \sqrt{n}\,\Delta x = S_n^* \sqrt{t}\,\frac{\Delta x}{\sqrt{\Delta t}} = S_n^* \sigma \sqrt{t},$$

because of (5.14) and (5.17). Applying Theorem 5.1, in particular formula (5.8), and passing to the limit $\Delta t \to 0$ while holding σ constant (so that $\Delta x \to 0$), we find that the random variable $\xi(t)$ describing the one-dimensional Brownian motion satisfies the formula

$$\mathbf{P}\left\{x' \leqslant \frac{\xi(t)}{\sigma\sqrt{t}} \leqslant x''\right\} = \lim_{\Delta t \to 0} \mathbf{P}\,\{x' \leqslant S_n^* \leqslant x''\} = \frac{1}{\sqrt{2\pi}} \int_{x'}^{x''} e^{-x^2/2}\, dx.$$

Therefore $\xi(t)$ is a normal random variable with probability distribution

$$\mathbf{P}\,\{x' \leqslant \xi(t) \leqslant x''\} = \frac{1}{\sigma\sqrt{2\pi t}} \int_{x'}^{x''} e^{-x^2/2\sigma^2 t}\, dx.$$

PROBLEMS

1. Consider the game of "heads or tails," as in Example 3, p. 29. Show that the probability of correctly calling the side of the coin landing upward is always $\frac{1}{2}$ regardless of the call, provided the coin is unbiased. However, show that if the coin is biased, then "heads" should be called all the time if heads are more likely, while "tails" should be called all the time if tails are more likely.

2. There are 10 children in a given family. Assuming that a boy is as likely to be born as a girl, what is the probability of the family having
a) 5 boys and 5 girls; b) From 3 to 7 boys?

3. Suppose the probability of hitting a target with a single shot is 0.001. What is the probability P of hitting the target 2 or more times in 5000 shots?

Ans. $P \approx 1 - 6e^{-5} \approx 0.96.$

4. The page proof of a 500-page book contains 500 misprints. What is the probability P of 2 or more misprints appearing on the same page?

Ans. $P \approx 1 - \dfrac{5}{2e} \approx 0.08.$

5. Let p be the probability of success in a series of Bernoulli trials. What is the probability P_n of an even number of successes in n trials?

Ans. $P_n = \frac{1}{2}[1 + (1 - 2p)^n].$

6. What is the probability of the pattern SFS appearing infinitely often in an infinite series of Bernoulli trials, if S denotes "success" and F "failure"?

Hint. Apply the second Borel-Cantelli lemma (Theorem 3.1, p. 33).[5]

Ans. 1.

7. An electronic computer contains 1000 transistors. Suppose each transistor has probability 0.001 of failing in the course of a year of operation. What is the probability of at least 3 transistors failing in a year?

8. A school has 730 students. What is the probability that exactly 4 students were born on January 1?

Hint. Neglect leap years.

9. Let ξ be a random variable with the Poisson distribution (5.6). Find

$$\text{a) } \sigma^2 = \mathbf{D}\xi; \quad \text{b) } \frac{\mathbf{E}(\xi - a)^3}{\sigma^3}.$$

Ans. a) a; b) $\dfrac{1}{\sqrt{a}}$.

10. Where is the uniform convergence used in the proof of Theorem 5.1?

11. The probability of occurrence of an event A in one trial is 0.3. What is the probability P that the relative frequency of A in 100 independent trials will lie between 0.2 and 0.4?

Hint. Use Theorem 5.1 and Table 2.

Ans. $P \approx 0.97$.

12. Suppose an event A has probability 0.4. How many trials must be performed to assert with probability 0.9 that the relative frequency of A differs from 0.4 by no more than 0.1?

Ans. About 65.

13. The probability of occurrence of an event A in one trial is 0.6. What is the probability P that A occurs in the majority of 60 trials?

Ans. $P \approx 0.94$.

14. Two continuous random variables ξ_1 and ξ_2 are said to have a *bivariate normal distribution* if their joint probability density is

$$p_{\xi_1,\xi_2}(x_1, x_2) = \frac{1}{2\pi\sigma_1\sigma_2\sqrt{1 - r^2}}$$
$$\times \exp\left\{-\frac{1}{2(1 - r^2)}\left[\frac{(x_1 - a)^2}{\sigma_1^2} - 2r\frac{(x_1 - a)(x_2 - b)}{\sigma_1\sigma_2} + \frac{(x_2 - b)^2}{\sigma_2^2}\right]\right\}, \tag{5.18}$$

---

[5] For further details, see W. Feller, *op. cit.*, p. 202.

where $\sigma_1 > 0$, $\sigma_2 > 0$, $-1 < r < 1$. Prove that each of the random variables ξ_1 and ξ_2 has a univariate (i.e., one-dimensional) normal distribution of the form (5.13), where $\mathbf{E}\xi_1 = a$, $\mathbf{D}\xi_1 = \sigma_1^2$, $\mathbf{E}\xi_2 = b$, $\mathbf{D}\xi_2 = \sigma_2^2$.

Hint. Clearly,

$$p_{\xi_1}(x_1) = \int_{-\infty}^{\infty} p_{\xi_1,\xi_2}(x_1, x_2)\,dx_2, \qquad p_{\xi_2}(x_2) = \int_{-\infty}^{\infty} p_{\xi_1,\xi_2}(x_1, x_2)\,dx_1$$

(why?).

15. Prove that the number r in (5.18) is the correlation coefficient of the random variables ξ_1 and ξ_2. Prove that ξ_1 and ξ_2 are independent if and only if $r = 0$.

Comment. This is a situation in which r is a satisfactory measure of the extent to which the random variables ξ_1 and ξ_2 are dependent (the larger $|r|$, the "more dependent" ξ_1 and ξ_2).

16. Let ξ_1 and ξ_2 be the same as in Problem 14. Find the probability distribution of $\eta = \xi_1 + \xi_2$.

Ans. The random variable η is normal, with probability density

$$p_\eta(x) = \frac{1}{\sqrt{2\pi(\sigma_1^2 + 2r\sigma_1\sigma_2 + \sigma_2^2)}} \exp\left[-\frac{(x - a - b)^2}{2(\sigma_1^2 + 2r\sigma_1\sigma_2 + \sigma_2^2)}\right].$$

6

SOME LIMIT THEOREMS

12. The Law of Large Numbers

Consider n independent identically distributed random variables $\xi_1, \ldots,$ ξ_n. In particular, $\xi_1, \ldots, \xi_n$ have the same mean $a = \mathbf{E}\xi_k$ and variance $\sigma^2 = \mathbf{D}\xi_k$. If

$$\eta = \frac{1}{n}(\xi_1 + \cdots + \xi_n)$$

is the arithmetic mean of the variables $\xi_1, \ldots, \xi_n$, then

$$\mathbf{E}\eta = \frac{1}{n}\sum_{k=1}^{n}\mathbf{E}\xi_k = a,$$

$$\mathbf{D}\eta = \mathbf{E}(\eta - a)^2 = \frac{1}{n^2}\sum_{k=1}^{n}\mathbf{D}\xi_k = \frac{\sigma^2}{n}.$$

Applying Chebyshev's inequality (see Sec. 9) to the random variable $\eta - a$, we get the inequality

$$\mathbf{P}\{|\eta - a| > \varepsilon\} \leqslant \frac{1}{\varepsilon^2}\mathbf{E}(\eta - a)^2 = \frac{\sigma^2}{n\varepsilon^2} \tag{6.1}$$

for arbitrary $\varepsilon > 0$.

THEOREM 6.1 (***Weak law of large numbers***). *Let $\xi_1, \ldots, \xi_n$ be n independent identically distributed random variables with mean a and variance σ^2. Then, given any $\delta > 0$ and $\varepsilon > 0$, however small, there is an integer* *n such that*

$$a - \varepsilon \leqslant \frac{1}{n}(\xi_1 + \cdots + \xi_n) \leqslant a + \varepsilon$$

with probability greater than $1 - \delta$.

Proof. The theorem is an immediate consequence of (6.1) if we choose $n > \sigma^2/\delta\varepsilon^2$. ▌

Remark. Suppose δ and ε are so small that we can practically neglect both the occurrence of events of probability δ and differences between quantities differing by no more than ε. Then Theorem 6.1 asserts that for sufficiently large n, the arithmetic mean

$$\eta = \frac{1}{n}(\xi_1 + \cdots + \xi_n)$$

is an excellent approximation to the mean value $a = \mathbf{E}\xi_k$

Now consider n consecutive Bernoulli trials, in each of which an event A can occur with probability $p = \mathbf{P}(A)$ or fail to occur with probability $q = 1 - p$. Let ξ_k be a random variable equal to 1 if A occurs at the kth trial and 0 if A fails to occur at the kth trial. Then the random variables $\xi_1, \ldots, \xi_n$ are independent and identically distributed (by the very meaning of Bernoulli trials). Obviously

$$\mathbf{P}\{\xi_k = 1\} = p, \qquad \mathbf{P}\{\xi_k = 0\} = q.$$

Moreover, each random variable ξ_k has mean

$$a = \mathbf{E}\xi_k = p \cdot 1 + q \cdot 0 = p = \mathbf{P}(A).$$

Let $n(A)$ be the number of trials in which A occurs, so that

$$\frac{n(A)}{n}$$

is the relative frequency of the event A. Then clearly

$$n(A) = \xi_1 + \cdots + \xi_n,$$

and hence

$$\frac{n(A)}{n} = \frac{1}{n}(\xi_1 + \cdots + \xi_n).$$

It follows from Theorem 6.1 that $n(A)/n$ virtually coincides with $\mathbf{P}(A)$ for sufficiently large n, more exactly, that given any $\delta > 0$ and $\varepsilon > 0$, however small, there is an integer n such that

$$\mathbf{P}(A) - \varepsilon \leqslant \frac{n(A)}{n} \leqslant \mathbf{P}(A) + \varepsilon$$

with probability greater than $1 - \delta$. The justification for formula (1.2), p. 3 is now apparent.

Remark. It can be shown[1] that with probability 1 the limit

$$\lim_{n\to\infty} \frac{n(A)}{A}$$

exists and equals $\mathbf{P}(A)$. This result is known as the *strong law of large numbers.*

13. Generating Functions. Weak Convergence of Probability Distributions

Let ξ be a discrete random variable taking the values $0, 1, 2, \ldots$ with probabilities

$$P_\xi(k) = \mathbf{P}\{\xi = k\}, \qquad k = 0, 1, 2, \ldots \tag{6.2}$$

Then the function

$$F_\xi(z) = \sum_{k=0}^{\infty} P_\xi(k) z^k, \qquad |z| \leqslant 1 \tag{6.3}$$

is called the *generating function* of the random variable ξ or of the corresponding probability distributions (6.2). It follows from the convergence of the series (6.3) for $|z| = 1$ and from Weierstrass's theorem on uniformly convergent series of analytic functions[2] that $F_\xi(z)$ is an analytic function of z in $|z| < 1$, with (6.3) as its power series expansion. Moreover, the probability distribution of the random variable ξ is uniquely determined by its generating function $F_\xi(z)$, and in fact

$$P_\xi(k) = \frac{1}{k!} F_\xi^{(k)}(0), \qquad k = 0, 1, 2, \ldots,$$

where $F_\xi^{(k)}(z)$ is the kth derivative of $F_\xi(z)$. According to formula (4.10), p. 44, for fixed z the function $F_\xi(z)$ is just the mathematical expectation of the random variable $\varphi(\xi) = z^\xi$, i.e.,

$$F_\xi(z) = \mathbf{E}z^\xi, \qquad |z| \leqslant 1. \tag{6.4}$$

***Example* 1 (*The Poisson distribution*).** If the random variable ξ has a Poisson distribution with parameter a, so that

$$P_\xi(k) = \frac{a^k}{k!} e^{-a}, \qquad k = 0, 1, 2, \ldots,$$

[1] See e.g., W. Feller, *op. cit.*, p. 203.

[2] See e.g., R. A. Silverman, *Introductory Complex Analysis*, Prentice-Hall, Inc., Englewood Cliffs, N.J. (1967), p. 191. Also use Weierstrass' *M*-test (*ibid.*, p. 186).

then ξ has the generating function

$$F_\xi(z) = \sum_{k=0}^{\infty} \frac{a^k}{k!} e^{-a} z^k = e^{-a} \sum_{k=0}^{\infty} \frac{(az)^k}{k!} = e^{a(z-1)}. \tag{6.5}$$

Suppose the random variable ξ has mean $a = \mathbf{E}\xi$ and variance $\sigma^2 = \mathbf{D}\xi$. Then, differentiating (6.4) twice with respect to z behind the expectation sign and setting $z = 1$, we get

$$a = \mathbf{E}\xi = F'(1), \qquad \sigma^2 = \mathbf{E}\xi^2 - (\mathbf{E}\xi)^2 = F''(1) + F'(1) - [F'(1)]^2. \tag{6.6}$$

The same formulas can easily be deduced from the power series (6.3). In fact, differentiating (6.3) for $|z| < 1$, we get

$$F'(z) = \sum_{k=0}^{\infty} kP_\xi(k) z^{k-1},$$

and hence

$$a = \sum_{k=0}^{\infty} kP_\xi(k) = \lim_{z \to 1} F'_\xi(z) = F'_\xi(1),$$

and similarly for the second of the formulas (6.6).

Next let $\xi_1, \ldots, \xi_n$ be n independent random variables taking the values $0, 1, 2, \ldots$ Then the random variables $z^{\xi_1}, \ldots, z^{\xi_n}$, where z is a fixed number, are also independent. It follows from formula (4.20), p. 47 that

$$\mathbf{E}z^{(\xi_1 + \cdots + \xi_n)} = \mathbf{E}z^{\xi_1} \cdots z^{\xi_n} = \mathbf{E}z^{\xi_1} \cdots \mathbf{E}z^{\xi_n}.$$

Thus we have the formula

$$F_\xi(z) = F_{\xi_1}(z) \cdots F_{\xi_n}(z), \tag{6.7}$$

expressing the generating function $F_\xi(z) = \mathbf{E}z^\xi$ of the sum $\xi = \xi_1 + \cdots + \xi_n$ of the n random variables $\xi_1, \ldots, \xi_n$ in terms of the generating functions $F_{\xi_k}(z) = \mathbf{E}z^{\xi_k}$, $k = 1, \ldots, n$ of the separate summands.

***Example* 2 (*The binomial distribution*).** Suppose the random variable ξ has a binomial distribution with parameters p and n, so that

$$P_\xi(k) = C_k^n p^k q^{n-k}, \qquad q = 1 - p, \quad k = 0, 1, \ldots, n.$$

Then, as already noted in Sec. 10, ξ can be regarded as the sum $\xi = \xi_1 + \cdots + \xi_n$ of n independent random variables $\xi_1, \ldots, \xi_n$, where

$$\xi_k = \begin{cases} 1 \text{ with probability } p, \\ 0 \text{ with probability } q. \end{cases}$$

The generating function $F_{\xi_k}(z)$ of each summand is clearly $pz + q$, and hence, by (6.7), the generating function of ξ itself is

$$F_\xi(z) = (pz + q)^n. \tag{6.8}$$

Now let ξ_n, $n = 1, 2, \ldots$ be a sequence of discrete random variables taking the values $0, 1, 2, \ldots$, with probability distributions $P_n(k) = P_{\xi_n}(k)$ and generating functions $F_n(z)$, $n = 1, 2, \ldots$ Then the sequence of distributions $\{P_n(k)\}$ is said to *converge weakly* to the limiting distribution $P(k)$ if

$$\lim_{n\to\infty} P_n(k) = P(k) \tag{6.9}$$

for all $k = 0, 1, 2, \ldots$

***Example* 3 (*Weak convergence of the binomial distribution to the Poisson distribution*).** Let $\xi_1, \xi_2, \ldots$ be a sequence of random variables such that ξ_n has a binomial distribution $P_n(k)$ with parameters p and n, i.e.,

$$P_n(k) = C_k^n p^k q^{n-k}, \qquad q = 1 - p.$$

Suppose p depends on n in such a way that the limit

$$\lim_{n\to\infty} np = a \tag{6.10}$$

exists. Then, according to formula (5.5), p. 55, the sequence of distributions $\{P_n(k)\}$ converges weakly to the Poisson distribution

$$P(k) = \frac{a^k}{k!} e^{-a}, \qquad k = 0, 1, 2, \ldots$$

with parameter a given by (6.10).

In Example 3, the sequence of generating functions

$$F_n(z) = (pz + q)^n, \qquad n = 1, 2, \ldots$$

of the random variables $\xi_1, \xi_2, \ldots$ converges uniformly to the generating function $F(z) = e^{a(z-1)}$ of the limiting Poisson distribution, i.e.,

$$\lim_{n\to\infty} F_n(z) = \lim_{n\to\infty} [1 + p(z-1)] = \lim_{n\to\infty} \left[1 + \frac{np(z-1)}{n}\right]^n$$
$$= \lim_{n\to\infty} \left[1 + \frac{a(z-1)}{n}\right]^n = e^{a(z-1)}$$

(justify the next-to-the-last step). This is no accident, as shown by

THEOREM 6.2. *The sequence of probability distributions $P_n(k)$, $n = 1, 2, \ldots$ with generating functions $F_n(z)$, $n = 1, 2, \ldots$ converges weakly to* *the limiting distribution $P(k)$ if and only if*

$$\lim_{n\to\infty} F_n(z) = F(z), \tag{6.11}$$

where

$$F(z) = \sum_{k=0}^{\infty} P(k)z^k$$

is the generating function of $P(k)$ and the convergence is uniform in every disk $|z| \leqslant r < 1$.

Proof. First suppose (6.9) holds. Clearly,

$$|F_n(z) - F(z)| \leqslant \sum_{k=0}^{K} |P_n(k) - P(k)| + \sum_{k=K+1}^{\infty} |z|^k \tag{6.12}$$

for any positive integer K. Given any $\varepsilon > 0$, we first choose K large enough to make

$$\sum_{k=K+1}^{\infty} |z|^k \leqslant \frac{r^{K+1}}{1-r} < \frac{\varepsilon}{2},$$

and then find a positive integer N such that

$$|P_n(k) - P(k)| < \frac{\varepsilon}{2(K+1)}$$

holds for $k = 0, \ldots, K$ if $n \geqslant N$. It then follows from (6.12) that

$$|F_n(z) - F(z)| < \varepsilon$$

if $n \geqslant N$, which immediately proves (6.11).

Conversely, suppose (6.11) holds, where the convergence is uniform in every disk $|z| \leqslant r < 1$. Then, by Weierstrass' theorem on uniformly convergent sequences of analytic functions,[3]

$$\lim_{n\to\infty} F_n^{(k)}(z) = F^{(k)}(z), \qquad |z| < 1 \tag{6.13}$$

for all $k = 0, 1, 2, \ldots$. But

$$P_n(k) = \frac{1}{k!} F_n^{(k)}(0), \qquad P(k) = \frac{1}{k!} F^{(k)}(0),$$

and hence (6.13) implies (6.9) for all $k = 0, 1, 2, \ldots$ ∎

The following example is typical of the situations where the Poisson distribution is encountered:

***Example* 4 (*Random flow of events*).** Suppose that events of a given kind occur randomly in the course of time. For example, we can think in terms

[3] R. A. Silverman, *op. cit.*, p. 192.

of "service calls" (requests for service) arriving randomly at some "server" (service facility), like inquiries at an information desk, arrival of motorists at a gas station, telephone calls at an exchange, etc. Let $\xi(\Delta)$ be the number of events occurring during the time interval Δ. Then what is the distribution of the random variable $\xi(\Delta)$?

To answer this question, we will assume that our "random flow of events" has the following three properties:

a) The events are independent of one another; more exactly, the random variables $\xi(\Delta_1)$, $\xi(\Delta_2)$, ... are independent if the intervals $\Delta_1, \Delta_2, \ldots$ are nonoverlapping.
b) The flow of events is "stationary," i.e., the distribution of the random variable $\xi(\Delta)$ depends only on the length of the interval Δ and not on the time of its occurrence (the initial time of Δ, say).
c) The probability that at least one event occurs in a small time interval Δt is $\lambda \Delta t + o(\Delta t)$, while the probability that more than one event occurs in Δt is $o(\Delta t)$. Here $o(\Delta t)$ is an infinitesimal of higher order than Δt, i.e.,

$$\lim_{\Delta t \to 0} \frac{o(\Delta t)}{\Delta t} = 0,$$

and λ is a positive parameter characterizing the "rate of occurrence" or "density" of the events.

Now consider the time interval $\Delta = [0, t]$, and let $\xi(t)$ be the total number of events occurring in $[0, t]$. Dividing $[0, t]$ into n equal parts $\Delta_1, \ldots, \Delta_n$, we find that

$$\xi(t) = \sum_{k=1}^{n} \xi(\Delta_k),$$

where $\xi(\Delta_1), \ldots, \xi(\Delta_n)$ are independent random variables and $\xi(\Delta_k)$ is the number of events occurring in the interval Δ_k. Clearly, the generating function of each random variable $\xi(\Delta_k)$ is

$$F_n(z) = \left(1 - \frac{\lambda t}{n}\right) + \frac{\lambda t}{n} z + o\left(\frac{t}{n}\right),$$

where $o(t/n)$ is a term of order higher than t/n. Hence, by (6.7), the generating function of $\xi(t)$ is

$$F(z) = [F_n(z)]^n = \left[1 + \frac{\lambda t(z-1)}{n} + o\left(\frac{t}{n}\right)\right]^n.$$

But $F(z)$ is independent of the subintervals $\Delta_1, \ldots, \Delta_n$, and hence we can take the limit as $n \to \infty$, obtaining

$$F(z) = \lim_{n \to \infty} \left[1 + \frac{\lambda t(z-1)}{n}\right]^n = e^{\lambda t(z-1)}.$$

Comparing this with (6.5), we find that $F(z)$ is the generating function of a Poisson distribution with parameter $a = \lambda t$, so that

$$\mathbf{P}\{\xi(t) = k\} = \frac{(\lambda t)^k}{k!} e^{-\lambda t}, \qquad k = 0, 1, 2, \ldots$$

Since

$$\lambda t = \mathbf{E}\xi(t),$$

the parameter λ is just the average number of events occurring per unit time.

14. Characteristic Functions. The Central Limit Theorem

Given a real random variable ξ, by the *characteristic function* of ξ is meant the function

$$f_\xi(t) = \mathbf{E}e^{i\xi t}, \qquad -\infty < t < \infty. \tag{6.14}$$

Clearly, $f_\xi(t)$ coincides for every fixed t with the mathematical expectation of the complex random variable $\eta = e^{i\xi t}$. For a discrete random variable taking the values $0, 1, 2, \ldots$, the characteristic function $f_\xi(t)$ coincides with the values of the generating function $F_\xi(z)$ on the boundary of the unit circle $|z| = 1$, i.e.,

$$f_\xi(t) = F_\xi(e^{it}) = \sum_{k=0}^{\infty} P_\xi(k)e^{ikt}.$$

This formula represents $f_\xi(t)$ as a *Fourier series*, with the probabilities $P_\xi(k) = \mathbf{P}\{\xi = k\}$, $k = 0, 1, 2, \ldots$ as its coefficients. Thus these probabilities $P_\xi(k)$ are uniquely determined by the characteristic function $f_\xi(t)$.

If ξ is a continuous random variable with probability density $p_\xi(x)$, then, by formula (4.18), p. 47, the characteristic function is the *Fourier transform* of the density $p_\xi(x)$:

$$f_\xi(t) = \int_{-\infty}^{\infty} e^{ixt} p_\xi(x)\, dx. \tag{6.15}$$

Inverting (6.15), we find that

$$p_\xi(x) = \frac{1}{2\pi} \int_{-\infty}^{\infty} e^{-ixt} f_\xi(t)\, dt, \tag{6.16}$$

at least at points where $p_\xi(x)$ is suitably well-behaved.[4] Thus $p_\xi(x)$ is uniquely determined by the characteristic function $f_\xi(t)$.

---

[4] If (6.16) fails, another inversion formula can be used, giving the distribution function $\Phi_\xi(x) = \mathbf{P}\{\xi \leqslant x\}$ in terms of the characteristic function $f_\xi(t)$ (see e.g., B. V. Gnedenko, *op. cit.*, Sec. 36). We can then deduce $p_\xi(x)$ from $\Phi_\xi(x)$ by differentiation, at least almost everywhere (recall footnote 2, p. 38).

***Example* 1.** Let ξ be a normally distributed random variable, with probability density

$$p(x) = \frac{1}{\sqrt{2\pi}} e^{-x^2/2}. \tag{6.17}$$

Then, by (6.15), the characteristic function of ξ is given by

$$\begin{aligned} f_\xi(t) &= \int_{-\infty}^{\infty} e^{ixt} p(x)\,dx = \frac{1}{\sqrt{2\pi}} \int_{-\infty}^{\infty} e^{ixt-(x^2/2)}\,dx \\ &= e^{-t^2/2} \frac{1}{\sqrt{2\pi}} \int_{-\infty}^{\infty} e^{-(x-it)^2/2}\,dx. \end{aligned} \tag{6.18}$$

The function $\varphi(z) = e^{-z^2/2}$ is an analytic function of the complex variable z, and hence, by Cauchy's integral theorem,[5] the integral of $\varphi(z)$ along the rectangular contour with vertices $(-N, 0)$, $(N, 0)$, $(N, -it)$, $(-N, -it)$ equals zero. Therefore

$$\begin{aligned} \frac{1}{\sqrt{2\pi}} \int_{-\infty}^{\infty} e^{-(x-it)^2/2}\,dx &= \lim_{N\to\infty} \frac{1}{\sqrt{2\pi}} \int_{-N}^{N} e^{-(x-it)^2/2}\,dx \\ &= \lim_{N\to\infty} \frac{1}{\sqrt{2\pi}} \int_{-N-it}^{N-it} e^{-z^2/2}\,dz = \lim_{N\to\infty} \frac{1}{\sqrt{2\pi}} \int_{-N}^{N} e^{-x^2/2}\,dx \\ &= \frac{1}{\sqrt{2\pi}} \int_{-\infty}^{\infty} e^{-x^2/2}\,dx, \end{aligned} \tag{6.19}$$

where we use the fact that the integral of $\varphi(z)$ along the vertical sides of the contour vanishes as $N \to \infty$ (why?). But

$$\frac{1}{\sqrt{2\pi}} \int_{-\infty}^{\infty} e^{-x^2/2}\,dx = \int_{-\infty}^{\infty} p(x)\,dx = 1,$$

as for any probability density. Hence (6.18) and (6.19) imply

$$f_\xi(t) = e^{-t^2/2}. \tag{6.20}$$

Now suppose the random variable ξ is such that $\mathbf{E}\,|\xi|^3$ exists. Then the characteristic function $f_\xi(t)$ has the expansion

$$f_\xi(t) = 1 + i\mathbf{E}\xi \cdot t - \frac{\mathbf{E}\xi^2}{2} t^2 + R(t), \tag{6.21}$$

where the remainder $R(t)$ satisfies the estimate

$$|R(t)| \leqslant C\mathbf{E}\,|\xi|^3 \cdot |t|^3$$

[5] R. A. Silverman, *op. cit.*, p. 146.

(C denotes a constant). In fact, we need only note that

$$e^{i\xi t} = 1 + i\xi t - \frac{\xi^2}{2} t^2 + \theta, \tag{6.22}$$

by Taylor's formula, where

$$|\theta| \leqslant C\,|\xi|^3\, t^3.$$

We then get (6.21) by taking the mathematical expectation of both sides of (6.22). In particular, it follows from (6.21) that the mean $a = \mathbf{E}\xi$ and variance $\sigma^2 = \mathbf{D}\xi$ are given by the formulas

$$a = -if_\xi'(0), \qquad \sigma^2 = -f_\xi''(0) + [f_\xi'(0)]^2. \tag{6.23}$$

***Example* 2.** According to (6.23), the normally distributed random variable ξ with probability density (6.17) has mean

$$a = -if'(0) = 0$$

and variance

$$\sigma^2 = -f''(0) = 1.$$

Formula (6.7) has a natural analogue for characteristic functions. In fact, if $\xi_1, \ldots, \xi_n$ are independent random variables with sum $\xi = \xi_1 + \cdots + \xi_n$, then, by formula (4.20), p. 47, the characteristic function of ξ is

$$f_\xi(t) = f_{\xi_1}(t) \cdots f_{\xi_n}(t). \tag{6.24}$$

Let ξ_n, $n = 1, 2, \ldots$ be a sequence of random variables with characteristic functions $f_n(t)$, $n = 1, 2, \ldots$ Then the sequence of probability distributions of $\xi_1, \xi_2, \ldots$ is said to *converge weakly* to the distribution with density $p(x)$ if

$$\lim_{n\to\infty} \mathbf{P}\,\{x' \leqslant \xi_n \leqslant x''\} = \int_{x'}^{x''} p(x)\,dx$$

for all x' and x'' ($x' \leqslant x''$). This should be compared with the definition of weak convergence for discrete random variables taking the values $0, 1, 2, \ldots$ given in Sec. 13.

Theorem 6.2 has a natural analogue for characteristic functions, whose proof will not be given here:[6]

THEOREM 6.2.′ *The sequence of probability distributions with characteristic functions $f_n(t)$, $n = 1, 2, \ldots$ converges weakly to the limiting distribution with density $p(x)$ if and only if*

$$\lim_{n\to\infty} f_n(t) = f(t), \tag{6.11′}$$

[6] For the proof, see e.g., B. V. Gnedenko, *op. cit.*, Sec. 38.

where

$$f(t) = \int_{-\infty}^{\infty} e^{ixt} p(x)\, dx$$

is the characteristic function of the limiting distribution and the convergence is uniform in every finite interval $t' \leqslant t \leqslant t''$.

We now prove a key proposition of probability theory, called the *central limit theorem*, which has the De Moivre-Laplace theorem (Theorem 5.1, p. 59) as a very special case. Roughly speaking, the central limit theorem asserts that the distribution of the sum of a large number of independent identically distributed random variables is approximately normal.

DEFINITION. *Given a sequence of random variables* ξ_k, $k = 1, 2, \ldots$ *with finite means* $a_k = \mathbf{E}\xi_k$ *and variances* $\sigma_k^2 = \mathbf{D}\xi_k$, *consider the "normalized sum"*

$$S_n^* = \frac{S_n - \mathbf{E}S_n}{\sqrt{\mathbf{D}S_n}},$$

where

$$S_n = \sum_{k=1}^{n} \xi_k.$$

Then the sequence ξ_k, $k = 1, 2, \ldots$ *is said to satisfy the* ***central limit theorem*** *if*[7]

$$\lim_{n\to\infty} \mathbf{P}\{x' \leqslant S_n^* \leqslant x''\} = \frac{1}{\sqrt{2\pi}} \int_{x'}^{x''} e^{-x^2/2}\, dx. \tag{6.25}$$

THEOREM 6.3. *Suppose the sequence of independent random variables* ξ_k, $k = 1, 2, \ldots$ *with means* a_k *and variances* σ_k^2 *satisfies the* ***Lyapunov condition***

$$\lim_{n\to\infty} \frac{1}{B_n^3} \sum_{k=1}^{n} \mathbf{E}\,|\xi_k - a_k|^3 = 0, \tag{6.26}$$

where

$$B_n^2 = \mathbf{D}S_n = \sum_{k=1}^{n} \sigma_k^2.$$

Then the sequence of random variables satisfies the central limit theorem.

Proof. Equation (6.25) means that the sequence of distributions of the normalized sums S_n^*, $n = 1, 2, \ldots$ converges weakly to the normal distribution with probability density (6.16). Hence, according to Theorem 6.2′, we need only show that the sequence of characteristic

---

[7] Cf. formula (5.8), p. 59. Note that the right-hand side of (6.25) equals $\Phi(x'') - \Phi(x')$, where $\Phi(x)$ is the distribution function of a normal random variable with mean 0 and variance 1.

functions $f_n(t)$, $n = 1, 2, \ldots$ of the random variables S_n^* converges uniformly in every finite interval $t' \leqslant t \leqslant t''$ to the characteristic function $f(t) = e^{-t^2/2}$ of this normal distribution (recall Example 1). Clearly,

$$S_n^* = \sum_{k=1}^{n} \frac{\xi_k - a_k}{B_n}, \qquad B_n^2 = \mathbf{D}S_n.$$

The random variable $\xi_k - a_k$ has zero mean and variance σ_k^2, and hence, by (6.21), has characteristic function

$$g_k(t) = 1 - \frac{\sigma_k^2}{2} t^2 + R_k(t),$$

where

$$|R_k(t)| \leqslant C\,|t|^3\,\mathbf{E}\,|\xi_k - a_k|^3$$

(C is some constant). Therefore the characteristic function of the random variable $\eta_k = (\xi_k - a_k)/B_n$ is

$$f_{kn}(t) = g_k\left(\frac{t}{B_n}\right) = 1 - \frac{\sigma_k^2}{2B_n^2} t^2 + R_k\left(\frac{t}{B_n}\right),$$

where

$$\left| R_k\left(\frac{t}{B_n}\right) \right| \leqslant C\,|t|^3 \frac{\mathbf{E}\,|\xi_k - a_k|^3}{B_n^3}.$$

It follows from (6.24) that the random variable $S_n^* = \eta_1 + \cdots + \eta_n$ has characteristic function

$$f_n(t) = \prod_{k=1}^{n} f_{kn}(t).$$

Hence

$$\ln f_n(t) = \sum_{k=1}^{n} f_{kn}(t) \sim \sum_{k=1}^{n} \left[-\frac{\sigma_k^2}{2B_n^2} t^2 + R_k\left(\frac{t}{B_n}\right) \right],$$

where, because of the hypothesis (6.26),

$$\left| \sum_{k=1}^{n} R_k\left(\frac{t}{B_n}\right) \right| \leqslant C\,|t|^3 \frac{1}{B_n^3} \sum_{k=1}^{n} \mathbf{E}\,|\xi_k - a_k|^3 \to 0$$

as $n \to \infty$ uniformly in every finite interval $t' \leqslant t \leqslant t''$. Therefore

$$\ln f_n(t) \sim -\frac{t^2}{2B_n^2} \sum_{k=1}^{n} \sigma_k^2 = -\frac{t^2}{2},$$

or equivalently

$$f_n(t) \to e^{-t^2/2}$$

as $n \to \infty$. ∎

***Example* 3.** The Lyapunov condition is always satisfied if the random variables $\xi_1, \xi_2, \ldots$ are identically distributed and if $\alpha = \mathbf{E}\,|\xi_k - a_k|^3$ exists. In fact,

$$B_n^2 = \sum_{k=1}^{n} \mathbf{D}\xi_k = n\sigma^2,$$

where $\sigma^2 = \mathbf{D}\xi_k$, and hence

$$\lim_{n\to\infty} \frac{1}{B_n^3} \sum_{k=1}^{n} \mathbf{E}\,|\xi_k - a_k|^3 = \lim_{n\to\infty} \frac{1}{\sqrt{n}} \frac{\alpha}{\sigma^3} = 0.$$

PROBLEMS

1. Show that the conclusion of Theorem 6.1 can be written in the form

$$\lim_{n\to\infty} \mathbf{P}\left\{\left|\frac{1}{n}\sum_{k=1}^{n} \xi_k - a\right| < \varepsilon\right\} = 1$$

for arbitrary $\varepsilon > 0$.

2. Let $\xi_1, \ldots, \xi_n$ be n independent identically distributed random variables, with common mean $a = \mathbf{E}\xi_k$ and variance $\xi^2 = \mathbf{D}\xi_k$. Suppose a is known. Can the quantity

$$\frac{1}{n}\sum_{k=1}^{n} (\xi_k - a)^2$$

be used to estimate σ^2?

3. A random variable ξ has probability density[8]

$$p_\xi(x) = \begin{cases} \dfrac{x^m}{m!}\, e^{-x} & \text{if } x \geqslant 0, \\ 0 & \text{otherwise,} \end{cases}$$

where m is a positive integer. Prove that

$$\mathbf{P}\,\{0 \leqslant \xi \leqslant 2(m+1)\} > \frac{m}{m+1}.$$

Hint. Use Chebyshev's inequality.

4. The probability of an event A occurring in one trial is $\frac{1}{2}$. Is it true that the probability of A occurring between 400 and 600 times in 1000 independent trials exceeds 0.97?

Ans. Yes.

5. Let ξ be the number of spots obtained in throwing an unbiased die. What is the generating function of ξ?

---

[8] It follows by repeated integration by parts that $\int_0^\infty x^m e^{-x}\,dx = m!$

6. Use (6.6) and the result of the preceding problem to solve Problem 16, p. 52.

7. Let ξ be a random variable with the Poisson distribution

$$P_\xi(k) = \frac{a^k}{k!} e^{-a}, \qquad k = 0, 1, 2, \ldots \tag{6.27}$$

Use (6.6) to show that $\mathbf{E}\xi = \mathbf{D}\xi = a$.

8. Find the generating function of the random variable ξ with distribution

$$\mathbf{P}\{\xi = k\} = \frac{a^k}{(1 + a)^{k+1}} \qquad (a > 0).$$

Use (6.6) to find $\mathbf{E}\xi$ and $\mathbf{D}\xi$.

9. Let η be the sum of two independent random variables ξ_1 and ξ_2, one with the Poisson distribution (6.27), the other with the Poisson distribution obtained by changing the parameter a to a' in (6.27). Show that η also has a Poisson distribution, with parameter $a + a'$.

10. Let S_n be the number of successes in a series of n independent trials, where the probability of success at the kth trial is p_k. Suppose $p_1, \ldots, p_n$ depend on n in such a way that

$$p_1 + \cdots + p_n = \lambda,$$

while

$$\max\{p_1, \ldots, p_n\} \to 0$$

as $n \to \infty$. Prove that S_n has a Poisson distribution with parameter λ in the limit as $n \to \infty$.

Hint. Use Theorem 6.2.[9]

11. Find the characteristic function $f_\xi(t)$ of the random variable with probability density

$$p_\xi(x) = \frac{1}{2} e^{-|x|} \qquad (-\infty < x < \infty).$$

Ans. $f_\xi(t) = \dfrac{1}{1 + t^2}.$

12. Use (6.23) and the result of the preceding problem to solve Problem 13, p. 52.

13. Find the characteristic function of a random variable uniformly distributed in the interval $[a, b]$.

14. A continuous random variable ξ has characteristic function

$$f_\xi(t) = e^{-a|t|} \qquad (a > 0).$$

Find the probability density of ξ.

---

[9] For the details, see W. Feller, *op. cit.*, p. 282.

Ans. $p_\xi(x) = \dfrac{a}{\pi(a^2 + x^2)}$.

15. The derivatives $f'_\xi(0)$ and $f''_\xi(0)$ do not exist in the preceding problem. Why does this make sense?

Hint. Cf. Problem 24, p. 53.

16. Let ν be the total number of spots which are obtained in 1000 independent throws of an unbiased die. Then $\mathbf{E}\nu = 3500$, because of Problem 16, p. 52. Estimate the probability that ν is a number between 3450 and 3550.

17. Let S_n be the same as in Problem 10, and suppose $\sum_{k=1}^{\infty} p_k q_k = \infty$. Prove that

$$\mathbf{P}\left\{x' \leqslant \frac{S_n - \sum_{k=1}^{n} p_k}{\sqrt{\sum_{k=1}^{n} p_k q_k}} \leqslant x''\right\} \to \frac{1}{\sqrt{2\pi}} \int_{x'}^{x''} e^{-x^2/2}\, dx$$

as $n \to \infty$.

Hint. Apply Theorem 6.3.

7

MARKOV CHAINS

15. Transition Probabilities

Consider a physical system with the following properties:

a) The system can occupy any of a finite or countably infinite number of states $\varepsilon_1, \varepsilon_2, \ldots$

b) Starting from some initial state at time $t = 0$, the system changes its state randomly at the times $t = 1, 2, \ldots$ Thus, if the random variable $\xi(t)$ is the state of the system at time t,[1] the evolution of the system in time is described by the consecutive transitions (or "steps")

$$\xi(0) \to \xi(1) \to \xi(2) \to \cdots.$$

c) At time $t = 0$, the system occupies the state ε_i with *initial probability*

$$p_i^0 = \mathbf{P}\{\xi(0) = \varepsilon_i\}, \qquad i = 1, 2, \ldots \tag{7.1}$$

d) Suppose the system is in the state ε_i at any time n. Then the probability that the system goes into the state ε_j at the next step is given by

$$p_{ij} = \mathbf{P}\{\xi(n+1) = \varepsilon_j \mid \xi(n) = \varepsilon_i\}, \qquad i, j = 1, 2, \ldots, \tag{7.2}$$

regardless of its behavior before the time n. The numbers p_{ij}, called the *transition probabilities*, do not depend on the time n.

[1] In calling $\xi(t)$ a random variable, we are tacitly assuming that the states $\varepsilon_1, \varepsilon_2, \ldots$ are numbers (random variables are numerical functions). This can always be achieved by the simple expedient of replacing $\varepsilon_1, \varepsilon_2, \ldots$ by the integers $1, 2, \ldots$ (see W. Feller, *op. cit.*, p. 419).

A "random process" described by this model is called a *Markov chain.*[2]

Now let

$$p_j(n) = \mathbf{P}\{\xi(n) = \varepsilon_j\} \tag{7.3}$$

be the probability that the system will be in the state ε_j "after n steps." To find $p_j(n)$, we argue as follows: After $n - 1$ steps, the system must be in one of the states ε_k, $k = 1, 2, \ldots$, i.e., the events $\{\xi(n-1) = \varepsilon_k\}$, $k = 1, 2, \ldots$ form a full set of mutually exclusive events in the sense of p. 26. Hence, by formula (3.6),

$$\mathbf{P}\{\xi(n) = \varepsilon_j\} = \sum_k \mathbf{P}\{\xi(n) = \varepsilon_j \mid \xi(n-1) = \varepsilon_k\}\, \mathbf{P}\{\xi(n-1) = \varepsilon_k\}. \tag{7.4}$$

Writing (7.4) in terms of the notation (7.1)–(7.3), we get the recursion formulas

$$\begin{aligned} p_j(0) &= p_j^0, \\ p_j(n) &= \sum_k p_k(n-1)p_{kj}, \qquad n = 1, 2, \ldots \end{aligned} \tag{7.5}$$

If the system is in a definite state ε_i at time $t = 0$, the initial probability distribution reduces to

$$p_i^0 = 1, \qquad p_k^0 = 0 \quad \text{if} \quad k \neq i. \tag{7.6}$$

The probability $p_j(n)$ is then the same as the probability

$$p_{ij}(n) = \mathbf{P}\{\xi(n) = \varepsilon_j \mid \xi(0) = \varepsilon_i\}, \quad i, j = 1, 2, \ldots$$

that the system will go from state ε_i to state ε_j in n steps. Hence, for the initial distribution (7.6), the formulas (7.5) become

$$\begin{aligned} p_{ij}(0) &= \begin{cases} 1 & \text{if} \quad j = i, \\ 0 & \text{if} \quad j \neq i, \end{cases} \\ p_{ij}(n) &= \sum_k p_{ik}(n-1)p_{kj}, \qquad n = 1, 2, \ldots \end{aligned} \tag{7.7}$$

The form of the sum in (7.7) suggests introducing the *transition probability matrix*

$$P = \|p_{ij}\| = \left\| \begin{matrix} p_{11} & p_{12} & \cdots \\ p_{21} & p_{22} & \cdots \\ \cdot & \cdot & \cdots \\ \cdot & \cdot & \cdots \end{matrix} \right\|$$

---

[2] More exactly, a *Markov chain with stationary transition probabilities*, where we allude to the fact that the numbers (7.2) do not depend on n. For an abstract definition of a Markov chain, without reference to an underlying physical system, see W. Feller, *op. cit.*, p. 374.

and the "n-step transition probability matrix"

$$P(n) = \|p_{ij}(n)\| = \left\| \begin{matrix} p_{11}(n) & p_{12}(n) & \cdots \\ p_{21}(n) & p_{22}(n) & \cdots \\ \cdot & \cdot & \cdots \\ \cdot & \cdot & \cdots \end{matrix} \right\|.$$

Then, because of the rule for matrix multiplication,[3] (7.7) implies

$$P(0) = I, \quad P(1) = P, \quad P(2) = P(1)P = P^2, \ldots,$$

where I is the unit matrix (with ones along the main diagonal and zeros everywhere else). It follows that

$$P(n) = P^n, \qquad n = 1, 2, \ldots \tag{7.8}$$

***Example* 1 (*The book pile problem*).** Consider a pile of m books lying on a desk. If the books are numbered from 1 to m, the order of the books from the top of the pile down is described by some permutation $(i_1, i_2, \ldots, i_m)$ of the integers $1, 2, \ldots, m$, where i_1 is the number of the book on top of the pile, i_2 the number of the next book down, etc., and i_m is the number of the book at the bottom of the pile. Suppose each of the books is chosen with a definite probability, and then returned to the top of the pile. Let p_k be the probability of choosing the kth book ($k = 1, 2, \ldots, m$), and suppose the book pile is in the state $(i_1, i_2, \ldots, i_m)$. Then, at the next step, the state either remains unchanged, which happens with probability p_{i_1} when the top book (numbered i_1) is chosen, or else changes to one of the $m - 1$ states of the form $(i_k, i_1, \ldots)$, which happens with probability p_{i_k} when a book other than the top book is chosen. Thus we are dealing with a Markov chain, with states described by the permutations $(i_1, i_2, \ldots, i_m)$ and the indicated transition probabilities.

For example, if $m = 2$, there are only two states $\varepsilon_1 = (1, 2)$ and $\varepsilon_2 = (2, 1)$, and the transition probabilities are

$$p_{11} = p_{21} = p_1, \qquad p_{12} = p_{22} = p_2.$$

The corresponding transition probability matrix is

$$P = \left\| \begin{matrix} p_1 & p_2 \\ p_1 & p_2 \end{matrix} \right\|.$$

The "two-step transition probabilities" are

$$p_{11}(2) = p_{21}(2) = p_1p_1 + p_1p_2 = p_1(p_1 + p_2) = p_1,$$
$$p_{12}(2) = p_{22}(2) = p_1p_2 + p_2p_2 = p_2(p_1 + p_2) = p_2.$$

---

[3] Suitably generalized to the case of infinite matrices, if there are infinitely many states $\varepsilon_1, \varepsilon_2, \ldots$

Hence $P^2 = P$, and more generally $P^n = P$. Given any initial probability distribution p_1^0, p_2^0, we have

$$p_1(n) = p_1^0 p_{11}(n) + p_2^0 p_{21}(n) = p_1(p_1^0 + p_2^0) = p_1,$$
$$p_2(n) = p_1^0 p_{12}(n) + p_2^0 p_{22}(n) = p_2(p_1^0 + p_2^0) = p_2.$$

***Example* 2 (*The optimal choice problem*).** Returning to Example 2, p. 28, concerning the choice of the best object among m objects all of different quality, let ε_k $(k = 1, 2, \ldots, m)$ be the state characterized by the fact that the kth inspected object is the best of the first k objects inspected, and let ε_{m+1} be the state characterized by the fact that the best of all m objects has already been examined and rejected. As the m objects are examined one by one at random, there are various times at which the last object examined turns out to be better than all previous objects examined. Denote these times in order of occurrence by $t = 0, 1, \ldots, \nu$, with $t = 0$ corresponding to inspection of the first object and $t = \nu$ being the time at which the best of all m objects is examined ($\nu = 0$ if the best object is examined first). Imagine a system with possible states $\varepsilon_1, \ldots, \varepsilon_m, \varepsilon_{m+1}$, and let $\xi(t)$ be the state of the system at the time t, so that in particular $\xi(0) = \varepsilon_1$. To make the "random process" $\xi(0) \to \xi(1) \to \xi(2) \to \cdots$ into a Markov chain, we must define $\xi(n)$ for $n > \nu$. This is done by the simple artifice of setting $\xi(n) = \varepsilon_{m+1}$ for all $n > \nu$.

The transition probabilities of this Markov chain are easily found. Obviously $p_{m+1,m+1} = 1$ and $p_{ij} = 0$ if $i \geqslant j$, $j \leqslant m$. To calculate p_{ij} for $i < j < m$, we write (7.2) in the form

$$p_{ij} = \mathbf{P}(E_j \mid E_i) = \frac{\mathbf{P}(E_i E_j)}{\mathbf{P}(E_i)}, \tag{7.9}$$

in terms of the events $E_i = \{\xi(n) = \varepsilon_i\}$ and $E_j = \{\xi(n+1) = \varepsilon_j\}$. Clearly, $\mathbf{P}(E_i)$ is the probability that the best object will occupy the last place in a randomly selected permutation of j objects, all of different quality. Since the total number of distinct permutations of j objects is $j!$, while the number of such permutations with a fixed element in the last (jth) place is $(j-1)!$, we have

$$\mathbf{P}(E_j) = \frac{(j-1)!}{j!} = \frac{1}{j}, \qquad j = 1, \ldots, m. \tag{7.10}$$

Similarly, $\mathbf{P}(E_i E_j)$ is the probability that the best object occupies the jth place, while a definite object (namely, the second best object) occupies the ith place. Clearly, there are $(j-2)!$ permutations of j objects with fixed elements in two places, and hence

$$\mathbf{P}(E_i E_j) = \frac{(j-2)!}{j!} = \frac{1}{(j-1)j}, \qquad i < j \leqslant m. \tag{7.11}$$

It follows from (7.7)–(7.11) that

$$p_{ij} = \frac{i}{(j-1)j}, \qquad i < j \leqslant m.$$

As for the transition probabilities $p_{i,m+1}$, they have in effect already been calculated in Example 2, p. 28:

$$p_{i,m+1} = \frac{i}{m}, \qquad i = 1, \ldots, m.$$

***Example* 3 (*One-dimensional random walk*).** Consider a particle which moves randomly along the x-axis, coming to rest only at the points $x = \ldots, -2, -1, 0, 1, 2, \ldots$ with integral coordinates. Suppose the particle's motion is such that once at a point i, it jumps at the next step to either the point $i + 1$ or the point $i - 1$, with probabilities p and $q = 1 - p$, respectively.[4] Let $\xi(n)$ be the particle's position after n steps. Then the sequence $\xi(0) \to \xi(1) \to \xi(2) \to \cdots$ is a Markov chain with transition probabilities

$$p_{ij} = \begin{cases} p & \text{if } j = i + 1, \\ q & \text{if } j = i - 1, \\ 0 & \text{otherwise.} \end{cases} \tag{7.12}$$

In another kind of one-dimensional random walk, the particle comes to rest only at the points $x = 0, 1, 2, \ldots$, jumping from the point i to the point $i + 1$ with probability p_i and returning to the origin with probability $q_i = 1 - p_i$. The corresponding Markov chain has transition probabilities

$$p_{ij} = \begin{cases} p_i & \text{if } j = i + 1, \\ q_i & \text{if } j = 0, \\ 0 & \text{otherwise.} \end{cases} \tag{7.13}$$

16. Persistent and Transient States

Consider a Markov chain with states $\varepsilon_1, \varepsilon_2, \ldots$ and transition probabilities p_{ij}, $i, j = 1, 2, \ldots$ Suppose the system is initially in the state ε_i. Let

$$u_n = p_{ii}(n),$$

and let v_n be the probability that the system returns to the initial state ε_i

---

[4] Thus the particle's motion is "generated" by an infinite sequence of Bernoulli trials (cf. the example on pp. 63–65, where $p = q = \frac{1}{2}$).

for the first time after precisely n steps. Then

$$u_n = u_0 v_n + u_1 v_{n-1} + \cdots + u_{n-1} v_1 + u_n v_0, \qquad n = 1, 2, \ldots, \tag{7.14}$$

where we set

$$u_0 = 1, \qquad v_0 = 0$$

by definition. To see this, let B_k ($k = 1, \ldots, n$) be the event that "the system returns to ε_i for the first time after k steps," B_{n+1} the event that "the system does not return at all to ε_i during the first n steps," and A the event that "the system is in the initial state ε_i after n steps." Then the events $B_1, \ldots, B_n, B_{n+1}$ form a full set of mutually exclusive events, and hence, by the "total probability formula" (3.6), p. 26,

$$\mathbf{P}(A) = \sum_{i=1}^{n+1} \mathbf{P}(A \mid B_k)\mathbf{P}(B_k), \tag{7.15}$$

where clearly $\mathbf{P}(A \mid B_{n+1}) = 0$ and

$$\mathbf{P}(B_k) = v_k, \quad \mathbf{P}(A \mid B_k) = u_{n-k}, \qquad k = 1, \ldots, n.$$

Substituting these values into (7.15), we get (7.14).

In terms of the generating functions[5]

$$U(z) = \sum_{k=0}^{\infty} u_k z^k, \qquad V(z) = \sum_{k=0}^{\infty} v_k z^k, \qquad |z| < 1,$$

we can write (7.14) in the form

$$U(z) - u_0 = U(z)V(z), \qquad u_0 = 1,$$

which implies

$$U(z) = \frac{1}{1 - V(z)}. \tag{7.16}$$

The quantity

$$v = \sum_{n=0}^{\infty} v_n$$

is the probability that the system sooner or later returns to the original state ε_i. The state ε_i is said to be *persistent* if $v = 1$ and *transient* if $v < 1$.

THEOREM 7.1. *The state ε_i is persistent if and only if*

$$\sum_{n=0}^{\infty} u_n = \sum_{n=0}^{\infty} p_{ii}(n) = \infty. \tag{7.17}$$

---

[5] Although the numbers $u_0, u_1, u_2, \ldots$ do not correspond to a probability distribution as on p. 70 (in fact, we will consider the case where $\sum_{k=0}^{\infty} u_k = \infty$), we continue to call $U(z)$ a "generating function." The convergence of the series $\sum_{k=0}^{\infty} u_k z^k$ for $|z| < 1$ follows by comparison with the geometric series, since $|u_k| \leqslant 1$ for every k.

Proof. To say that ε_i is persistent means that

$$v = \sum_{n=0}^{\infty} v_n = \lim_{z \to 1} V(z) = 1,$$

or equivalently,

$$\lim_{z \to 1} U(z) = \lim_{z \to 1} \frac{1}{1 - V(z)} = \infty.$$

Suppose

$$\sum_{n=0}^{\infty} u_n < \infty. \tag{7.18}$$

Then, since the u_n are all nonnegative,

$$\sum_{n=0}^{N} u_n \leqslant \lim_{z \to 1} U(z) \leqslant \sum_{n=0}^{\infty} u_n$$

for every N, and hence, taking the limit as $N \to \infty$, we have

$$\lim_{z \to 1} U(z) = \sum_{n=0}^{\infty} u_n.$$

In other words, $U(z)$ approaches a finite limit as $z \to 1$ if and only if (7.18) holds. Equivalently, $U(z) \to \infty$ as $z \to 1$, i.e., ε_i is persistent, if and only if (7.17) holds. ∎

THEOREM 7.2. *If the initial state ε_i is persistent, then with probability 1 the system returns infinitely often to ε_i as the number of steps $n \to \infty$. If ε_i is transient, then with probability 1 the system returns to ε_i only finitely often, i.e., after a certain number of steps the system never again returns to ε_i.*

Proof. Suppose the system first returns to ε_i after ν_1 steps, returns a second time to ε_i after ν_2 steps, and so on. If there are fewer than k returns to ε_i as $n \to \infty$, we set $\nu_k = \infty$. Then the event $\{\nu_k < \infty\}$ means that there are at least k returns to ε_i, and the probability of the system returning to ε_i at least once is just

$$\mathbf{P}\{\nu_1 < \infty\} = v.$$

If the event $\{\nu_1 < \infty\}$ occurs, the system returns to its initial state ε_i after ν_1 steps, and its subsequent behavior is the same as if it just started its motion in ε_1. It follows that

$$\mathbf{P}\{\nu_2 < \infty \mid \nu_1 < \infty\} = v.$$

Clearly $\nu_1 = \infty$ implies $\nu_2 = \infty$, and hence $\nu_2 < \infty$ implies $\nu_1 < \infty$. Therefore

$$\mathbf{P}\{\nu_2 < \infty\} = \mathbf{P}\{\nu_2 < \infty \mid \nu_1 < \infty\}\mathbf{P}\{\nu_1 < \infty\} = v^2,$$

and similarly,

$$\mathbf{P}\{\nu_k < \infty \mid \nu_{k-1} < \infty\} = v, \quad \mathbf{P}\{\nu_k < \infty\} = v^k.$$

If ε_i is transient, then $v < 1$ and hence

$$\sum_{k=1}^{\infty} \mathbf{P}\{\nu_k < \infty\} = \sum_{k=1}^{\infty} v^k < \infty.$$

Therefore, by the first Borel-Cantelli lemma (Theorem 2.5, p. 21), with probability 1 only finitely many of the events $\{\nu_k < \infty\}$ occur, i.e., with probability 1 the system returns to the state ε_i only finitely often. This proves the second assertion in the statement of the theorem.

On the other hand, if ε_i is persistent, then $v = 1$, which implies

$$\mathbf{P}\{\nu_k < \infty\} = 1$$

for every k. Let $\varkappa$ be the number of times the system returns to its initial state ε_i as $n \to \infty$. Then obviously the events $\{\varkappa \geqslant k\}$ and $\{\nu_k < \infty\}$ are equivalent, so that if $\mathbf{P}\{\nu_k < \infty\} = 1$ for every k, then $\varkappa$ exceeds any preassigned integer k with probability 1. But then

$$\mathbf{P}\{\varkappa = \infty\} = 1,$$

which proves the first assertion. ∎

A state ε_j is said to be *accessible* from a state ε_i if the probability of the system going from ε_i to ε_j in some number of steps is positive, i.e., if $p_{ij}(M) > 0$ for some M.

THEOREM 7.3. *If a state ε_j is accessible from a persistent state ε_i, then ε_i is in turn accessible from ε_j and ε_j is itself persistent.*

Proof. Suppose ε_i is not accessible from ε_j. Then the system will go from ε_i to ε_j with positive probability $p_{ij}(M) = \alpha > 0$ for some number of steps M, after which the system cannot return to ε_i. But then the probability of the system eventually returning to ε_i cannot exceed $1 - \alpha$, contrary to the assumption that ε_i is persistent. Hence ε_i must be accessible from ε_j, i.e., $p_{ji}(N) = \beta > 0$ for some N. It follows from (7.8) that

$$P(n + M + N) = P(M)P(n)P(N) = P(N)P(n)P(M),$$

and hence

$$p_{ii}(n + M + N) \geqslant p_{ij}(M)p_{jj}(n)p_{ji}(N) = \alpha\beta p_{jj}(n),$$
$$p_{jj}(n + M + N) \geqslant p_{ji}(N)p_{ii}(n)p_{ij}(N) = \alpha\beta p_{ii}(n).$$

These inequalities show that the series

$$\sum_{n=0}^{\infty} p_{ii}(n), \qquad \sum_{n=0}^{\infty} p_{jj}(n)$$

either both converge or both diverge. But

$$\sum_{n=0}^{\infty} p_{ii}(n) = \infty$$

by Theorem 7.1, since ε_i is persistent. Therefore

$$\sum_{n=0}^{\infty} p_{jj}(n) = \infty,$$

i.e., ε_j is also persistent (again by Theorem 7.1). ∎

COROLLARY. *If a Markov chain has only a finite number of states, each accessible from every other state, then the states are all persistent.*

Proof. Since there are only a finite number of states, the system must return to at least one of them infinitely often as $n \to \infty$. Hence at least one of the states, say ε_i, is persistent. But all the other states are accessible from ε_i. It follows from Theorem 7.3 that all the states are persistent. ∎

***Example* 1.** In the book pile problem (Example 1, p. 85), if every book is chosen with positive probability, i.e., if $p_i > 0$ for all $i = 1, \ldots, m$, then obviously every state is accessible from every other state. In this case, all $m!$ distinct states $(i_1, \ldots, i_m)$ are persistent. If $p_i = 0$ for some i, then all states of the form $(i_1, \ldots, i_m)$ where $i_1 = i$ (the ith book lies on top of the pile) are transient, since at the very first step a book with a number j different from i will be chosen, and then the book numbered i, which can never be chosen from the pile, will steadily work its way downward.

***Example* 2.** In the optimal choice problem (Example 2, p. 86), it is obvious that after no more than m steps (m is the total number of objects), the system will arrive at the state ε_{m+1}, where it will remain forever. Hence all the states except ε_{m+1} are transient.

***Example* 3.** Consider the one-dimensional random walk with transition probabilities (7.12). Clearly, every state (i.e., every position of the particle) is accessible from every other state, and moreover[6]

$$p_{ii}(k) = \begin{cases} 0 & \text{if} \quad k = 2n + 1, \\ C_n^{2n} p^n q^n & \text{if} \quad k = 2n. \end{cases}$$

Using Stirling's formula (see p. 10), we have

$$C_n^{2n} p^n q^n = \frac{(2n)!}{(n!)^2} p^n q^n \sim \frac{\sqrt{4\pi n}\,(2n)^{2n} e^{-2n}}{(\sqrt{2\pi n}\, n^n e^{-n})^2} p^n q^n = \frac{1}{\sqrt{\pi n}} (4pq)^n$$

---

[6] Cf. formula (5.2), p. 55.

for large n, where

$$4pq = (p+q)^2 - (p-q)^2 = 1 - (p-q)^2 \leqslant 1$$

(the equality holds only for $p = q = \frac{1}{2}$). Therefore

$$p_{ii}(2n) \sim \frac{1}{\sqrt{\pi n}}(4pq)^n$$

for large n, and hence the series

$$\sum_{n=0}^{\infty} p_{ii}(2n), \qquad \sum_{n=0}^{\infty} \frac{1}{\sqrt{\pi n}}(4pq)^n$$

either both converge or both diverge. Suppose $p \neq q$, so that $4pq < 1$. Then

$$\sum_{n=0}^{\infty} p_{ii}(2n) < \infty,$$

and hence every state is transient. It is intuitively clear that if $p > q$ (say), then the particle will gradually work its way out along the x axis in the positive direction, and sooner or later permanently abandon any given state i. However, if $p = q = \frac{1}{2}$, we have

$$\sum_{n=0}^{\infty} p_{ii}(2n) = \infty,$$

and the particle will return to each state infinitely often, a fact apparent from the symmetry of the problem in this case.

***Example* 4.** Next consider the one-dimensional random walk with transition probabilities (7.13). Obviously, if $0 < p_i < 1$ for all $i = 0, 1, \ldots,$ every state is accessible from every other state, and hence the states are either all persistent or all transient. Suppose the system is initially in the state $i = 0$. Then the probability that it does not return to the state $i = 0$ after n steps equals the product $p_0 p_1 \cdots p_{n-1}$, the probability of the system making the consecutive transitions $0 \to 1 \to \cdots \to n$. It is easy to see that the probability that the system never returns to its initial state $i = 0$ as $n \to \infty$ equals the infinite product

$$\prod_{n=0}^{\infty} p_n = \lim_{n\to\infty} p_0 p_1 \cdots p_n.$$

If this infinite product converges to zero, i.e., if

$$\lim_{n\to\infty} p_0 p_1 \cdots p_n = 0,$$

then the state $i = 0$ is persistent, and hence so are all the other states. Otherwise, the probability of return to the initial state is

$$v = 1 - \lim_{n\to\infty} p_0 p_1 \cdots p_n < 1. \tag{7.19}$$

Then the state $i = 0$ is transient, and hence so are all the other states.

We can arrive at the same result somewhat differently by direct calculation of the probability v_n that the particle first returns to its initial state $i = 0$ in precisely n steps. Obviously, v_n is just the probability of the particle making the consecutive transitions $0 \to 1 \to \cdots \to n - 1$ in the first $n - 1$ steps and then returning to the state $i = 0$ in the nth step. Therefore, since the transition $i - 1 \to i$ has probability p_{i-1},

$$v_1 = 1 - p_0,$$
$$v_n = p_0 p_1 \cdots p_{n-2}(1 - p_{n-1}), \qquad n = 2, 3, \ldots$$

By definition, the probability of eventually returning to the initial state $i = 0$ is

$$v = \sum_{n=0}^{\infty} v_n.$$

Therefore

$$v = 1 - p_0 + p_0(1 - p_1) + p_0 p_1(1 - p_2) + \cdots = 1 - \lim_{n\to\infty} p_0 p_1 \cdots p_n,$$

in keeping with (7.19).

17. Limiting Probabilities. Stationary Distributions

As before, let $p_j(n)$ be the probability of the system occupying the state ε_j after n steps. Then, under certain conditions, the numbers $p_j(n)$, $j = 1, 2, \ldots$ approach definite limits as $n \to \infty$:

THEOREM 7.4. *Given a Markov chain with a finite number of states $\varepsilon_1, \ldots, \varepsilon_m$, each accessible from every other state, suppose*

$$\min_{i,j} p_{ij}(N) = \delta > 0 \tag{7.20}$$

for some N.[7] *Then*

$$\lim_{n\to\infty} p_j(n) = p_j^*,$$

where the numbers p_j^*, $j = 1, \ldots, m$, called the ***limiting probabilities,***[8] *do not depend on the initial probability distribution and satisfy the inequalities*

$$\max_i |p_{ij}(n) - p_j^*| \leqslant Ce^{-Dn}, \qquad |p_j(n) - p_j^*| \leqslant Ce^{-Dn} \tag{7.21}$$

for suitable positive constants C and D.

Proof. Let

$$r_j(n) = \min_i p_{ij}(n), \qquad R_j(n) = \max_i p_{ij}(n).$$

---

[7] In other words, suppose the probability of the system going from any state ε_i to any other state ε_j in some (fixed) number of steps N is positive.

[8] Clearly, the numbers p_j^* are nonnegative and have the sum 1 (why?). Hence they are candidates for the probabilities of a discrete probability distribution, as implicit in the term "limiting probabilities."

Then

$$r_j(n+1) = \min_i p_{ij}(n+1) = \min_i \sum_{k=1}^{m} p_{ik}p_{kj}(n) \geqslant \min_i \sum_{k=1}^{m} p_{ik}r_j(n) = r_j(n),$$

$$R_j(n+1) = \max_i p_{ij}(n+1) = \max_i \sum_{k=1}^{m} p_{ik}p_{kj}(n) \leqslant \max_i \sum_{k=1}^{m} p_{ik}R_j(n) = R_j(n),$$

and hence

$$r_j(1) \leqslant r_j(2) \leqslant \cdots \leqslant r_j(n) \leqslant \cdots \leqslant R_j(n) \leqslant \cdots \leqslant R_j(2) \leqslant R_j(1).$$

Let N be the same as in (7.20). Then, for arbitrary states ε_α and ε_β,

$$\sum_{k=1}^{m} p_{\alpha k}(N) = \sum_{k=1}^{m} p_{\beta k}(N) = 1.$$

Therefore

$$\sum_{k=1}^{m} p_{\alpha k}(N) - \sum_{k=1}^{m} p_{\beta k}(N)$$
$$= \sum_k{}^{+}[p_{\alpha k}(N) - p_{\beta k}(N)] + \sum_k{}^{-}[p_{\alpha k}(N) - p_{\beta k}(N)] = 0,$$

where the sum $\sum^{+}$ ranges over all k such that $p_{\alpha k}(N) - p_{\beta k}(N) > 0$ and $\sum^{-}$ ranges over all k such that $p_{\alpha k}(N) - p_{\beta k}(N) < 0$. Clearly, (7.20) implies

$$\max_{\alpha,\beta} \sum_k{}^{+} [p_{\alpha k}(N) - p_{\beta k}(N)] = d < 1,$$

for some positive number d.

Next we estimate the differences $R_j(n) - r_j(n)$ and $R_j(n+N) - r_j(n+N)$:

$$\begin{aligned}
R_j(N) - r_j(N) &= \max_\alpha p_{\alpha j}(N) - \min_\beta p_{\beta j}(N) \\
&= \max_{\alpha,\beta} [p_{\alpha j}(N) - p_{\beta j}(N)] \\
&\leqslant \max_{\alpha,\beta} \sum_k{}^{+} [p_{\alpha k}(N) - p_{\beta k}(N)] = d, \\
R_j(n+N) - r_j(n+N) &= \max_{\alpha,\beta} [p_{\alpha j}(n+N) - p_{\beta j}(n+N)] \\
&= \max_{\alpha,\beta} \sum_{k=1}^{m} [p_{\alpha k}(N) - p_{\beta k}(N)]p_{kj}(n) \\
&\leqslant \max_{\alpha,\beta} \Big\{\sum_k{}^{+} [p_{\alpha k}(N) - p_{\beta k}(N)]R_j(n) \\
&\qquad + \sum_k{}^{-} [p_{\alpha k}(N) - p_{\beta k}(N)]r_j(n)\Big\} \\
&= \max_{\alpha,\beta} \Big\{\sum_k{}^{+} [p_{\alpha k}(N) - p_{\beta k}(N)][R_j(n) - r_j(n)]\Big\} \\
&= d[R_j(n) - r_j(n)].
\end{aligned}$$

It follows that

$$R_j(kN) - r_j(kN) \leqslant d^k, \qquad k = 1, 2, \ldots \tag{7.22}$$

But, as already noted, the sequence $r_j(n)$, $n = 1, 2, \ldots$ is nondecreasing while the sequence $R_j(n)$, $n = 1, 2, \ldots$ is nonincreasing, and moreover $r_j(n) \leqslant R_j(n)$. Hence (7.22) shows that both sequences have the same limit

$$p_j^* = \lim_{n\to\infty} r_j(n) = \lim_{n\to\infty} R_j(n).$$

Moreover, it is clear that

$$|p_{ij}(n) - p_j^*| \leqslant R_j(n) - r_j(n) \leqslant d^{(n/N)-1}, \qquad i = 1, \ldots, n. \tag{7.23}$$

Therefore, given any initial distribution p_i^0, $i = 1, \ldots, n$, we have

$$\begin{aligned} |p_j(n) - p_j^*| &= \left| \sum_{i=1}^m p_i^0 p_{ij}(n) - p_j^* \right| = \left| \sum_{i=1}^m p_i^0 [p_{ij}(n) - p_j^*] \right| \\ &\leqslant \sum_{i=1}^m p_i^0 [R_j(n) - r_j(n)] = R_j(n) - r_j(n) \leqslant d^{(n/N)-1}, \qquad d < 1. \end{aligned} \tag{7.24}$$

But then

$$\lim_{n\to\infty} |p_j(n) - p_j^*| = 0,$$

i.e.,

$$\lim_{n\to\infty} p_j(n) = p_j^*$$

independently of the initial distribution, as asserted. Choosing

$$C = \frac{1}{d}, \qquad D = -\frac{1}{N} \ln d$$

in (7.23) and (7.24), we get (7.21). ▮

COROLLARY. *The limiting probabilities p_j^*, $j = 1, \ldots, m$ are a solution of the system of linear equations*

$$p_j^* = \sum_{i=1}^m p_i^* p_{ij}, \qquad j = 1, \ldots, m. \tag{7.25}$$

Proof. According to (7.5),

$$p_j(n) = \sum_k p_i(n-1) p_{ij}.$$

But this becomes (7.25), after taking the limit as $n \to \infty$. ▮

Remark. Given an arbitrary Markov chain with states $\varepsilon_1, \varepsilon_2, \ldots$, let p_i^0, $i = 1, 2, \ldots$ be numbers such that

$$p_i^0 \geqslant 0, \qquad \sum_i p_i^0 = 1$$

and

$$p_j^0 = \sum_i p_i^0 p_{ij}, \qquad j = 1, 2, \ldots \tag{7.26}$$

Choosing p_i^0, $i = 1, 2, \ldots$ as the initial probability distribution, we calculate the probability $p_j(n)$ of finding the system in the state ε_j after n steps, obtaining

$$p_j(1) = \sum_i p_i^0 p_{ij} = p_j^0,$$

$$p_j(2) = \sum_i p_i(1) p_{ij} = \sum_i p_i^0 p_{ij} = p_j^0,$$

$$\cdots\cdots\cdots\cdots$$

It follows that

$$p_j(n) = p_j^0, \qquad j = 1, 2, \ldots \tag{7.27}$$

for all $n = 0, 1, 2, \ldots$ [$p_j(0) = p_j^0$ trivially], i.e., the probabilities $p_j(n)$, $j = 1, 2, \ldots$ remain unchanged as the system evolves in time.

A Markov chain is said to be *stationary* if the probabilities $p_j(n)$, $j = 1, 2, \ldots$ remain unchanged for all $n = 0, 1, 2, \ldots$, and then the corresponding probability distribution with probabilities (7.27) is also said to be *stationary*. It follows from the corollary and the remark that a probability distribution p_j^0, $j = 1, 2, \ldots$ is stationary if and only if it satisfies the system of equations (7.26). Moreover, if the limiting probabilities

$$p_j^* = \lim_{n \to \infty} p_j(n) \tag{7.28}$$

are the same for every initial distribution, then there is a unique stationary distribution with probabilities

$$p_j^0 = p_j^*, \qquad j = 1, 2, \ldots$$

Hence Theorem 7.4 and its corollary can be paraphrased as follows: *Subject to the condition* (7.20), *the limiting probabilities* (7.28) *exist and are the unique solution of the system of linear equations* (7.25) *satisfying the extra conditions*

$$p_j^* \geqslant 0, \qquad \sum_{j=1}^{m} p_j^* = 1.$$

Moreover, they form a stationary distribution for the given Markov chain.

***Example* 1.** In the book pile problem, it will be recalled from p. 86 that when $m = 2$, the stationary distribution

$$p_1(n) = p_1, \qquad p_2(n) = p_2$$

is established at the very first step. In the case of arbitrary m, let $p_{(i_1,\ldots,i_m),(j_1,\ldots,j_m)}$ denote the probability of the transition from the state $(i_1, \ldots, i_m)$ to the state $(j_1, \ldots, j_m)$, and assume that the probabilities $p_1, \ldots, p_m$ are all positive. Then, as shown on p. 85,

$$p_{(i_1,\ldots,i_m),(j_1,\ldots,j_m)} = \begin{cases} p_{i_k} & \text{if } (j_1, \ldots, j_m) = (i_k, i_1, \ldots), \\ 0 & \text{otherwise,} \end{cases}$$

where the permutation $(i_k, i_1, \ldots)$ is obtained from $(i_1, \ldots, i_m)$ by choosing some i_k and moving it into the first position. The limiting probabilities $p^*_{(j_1,\ldots,j_m)}$ are the solution of the system of linear equations

$$p^*_{(j_1,\ldots,j_m)} = p_{j_1} \sum_{(j_1',\ldots,j_m')} p^*_{(j_1',\ldots,j_m')}, \tag{7.29}$$

where $(j_1', \ldots, j_m')$ ranges over the m permutations

$$(j_1, j_2, j_3, \ldots, j_m), (j_2, j_1, j_3, \ldots, j_m), \ldots, (j_1, j_2, \ldots, j_m, j_1)$$

which give $(j_1, \ldots, j_m)$ when j_1 is moved into the first position.

After a sufficiently large number of steps, a stationary distribution will be virtually established, i.e., the book pile will occupy the states $(i_1, \ldots, i_m)$ with virtually unchanging probabilities $p^*_{(i_1,\ldots,i_m)}$. Clearly, the probability of finding the ith book on top of the pile is then

$$p_i^* = \sum_{i_2,\ldots,i_m} p^*_{(i,i_2,\ldots,i_m)},$$

and hence, by (7.29),

$$p_i^* = \sum_{i_2,\ldots,i_m} p_i \sum_{(i_1',\ldots,i_m')} p^*_{(i_1',\ldots,i_m')},$$

where $(i_1', \ldots, i_m)$ ranges over the m permutations

$$(i, i_2, i_3, \ldots, i_m), (i_2, i, i_3, \ldots, i_m), \ldots, (i_2, i_3, \ldots, i_m, i)$$

which give $(i, i_2, \ldots, i_m)$ when i is moved into the first position. But then

$$p_i^* = p_i \sum_{i_1,\ldots,i_m} p^*_{(i_1,\ldots,i_m)} = p_i, \qquad i = 1, \ldots, m,$$

i.e., the limiting probability p_i^* of finding the ith book on top of the pile is just the probability p_i with which the ith book is chosen. Thus, the more often a book is chosen, the greater the probability of its ending up on top of the pile (which is hardly surprising!).

***Example* 2.** Consider the one-dimensional random walk with transition probabilities (7.12). If $p \neq q$, then the particle gradually moves further and further away from the origin, in the positive direction if $p > q$ and in the negative direction if $p < q$. If $p = q$, the particle will return infinitely often to each state, but for any fixed j, the probability $p_j(n)$ of the particle being at the point j approaches 0 as $n \to \infty$ (why?). Hence, in any case,

$$\lim_{n\to\infty} p_j(n) = p_j^* = 0$$

for every j, but the numbers p_j^*, $j = 1, 2, \ldots$ cannot be interpreted as the limiting probabilities, since they are all zero. In particular, there is no stationary distribution.

Example 3. Finally, consider the one-dimensional random walk with transition probabilities (7.13). Suppose

$$\lim_{n\to\infty} p_0 p_1 \cdots p_n = 1 - v > 0, \tag{7.30}$$

so that the states are all transient (see p. 92). Then as $n \to \infty$, the particle "moves off to infinity" in the positive direction with probability 1, and there is obviously no stationary distribution. If there is a stationary distribution, it must satisfy the system of equations (7.26), which in the present case take the form

$$p_j^0 = p_{j-1}^0 p_{j-1}, \qquad j = 1, 2, \ldots \tag{7.31}$$

It follows from (7.31) that

$$p_1^0 = p_0^0 p_0, \qquad p_2^0 = p_0^0 p_0 p_1, \ldots, \quad p_n^0 = p_0^0 p_0 p_1 \cdots p_{n-1}, \ldots$$

Clearly a stationary distribution exists if and only if the series

$$\sum_{n=0}^{\infty} p_0 p_1 \cdots p_n = 1 + p_0 + p_0 p_1 + \cdots \tag{7.32}$$

converges.[9] The stationary distribution is then

$$p_0^0 = \frac{1}{1 + p_0 + p_0 p_1 + \cdots},$$

$$p_n^0 = \frac{p_0 p_1 \cdots p_n}{1 + p_0 + p_0 p_1 + \cdots}, \qquad n = 1, 2, \ldots$$

PROBLEMS

1. A number from 1 to m is chosen at random, at each of the times $t = 1, 2, \ldots$ A system is said to be in the state ε_0 if no number has yet been chosen, and in the state ε_i if the largest number so far chosen is i. Show that the random process described by this model is a Markov chain. Find the corresponding transition probabilities p_{ij} $(i, j = 0, 1, \ldots, m)$.

Ans. $p_{ii} = \dfrac{i}{m}, \quad p_{ij} = 0$ if $i > j$, $\quad p_{ij} = \dfrac{1}{m}$ if $i < j$.

2. In the preceding problem, which states are persistent and which transient?

3. Suppose $m = 4$ in Problem 1. Find the matrix $P(2) = \|p_{ij}(2)\|$, where $p_{ij}(2)$ is the probability that the system will go from state ε_i to state ε_j in 2 steps.

---

[9] Note that (7.32) automatically diverges if (7.30) holds.

4. An urn contains a total of N balls, some black and some white. Samples are drawn from the urn, m balls at a time ($m \leqslant N$). After drawing each sample, the black balls are returned to the urn, while the white balls are replaced by black balls and then returned to the urn. If the number of white balls in the urn is i, we say that the "system" is in the state ε_i. Prove that the random process described by this model is a Markov chain (imagine that samples are drawn at the times $t = 1, 2, \ldots$ and that the system has some initial probability distribution). Find the corresponding transition probabilities p_{ij} ($i, j = 0, 1, \ldots, N$). Which states are persistent and which transient?

Ans. $p_{ij} = 0$ if $i < j$ or if $i \geqslant j$, $j > N - m$,

$$p_{ij} = \frac{C_{i-j}^{i} C_{m-i+j}^{N-i}}{C_m^N} \text{ if } i \geqslant j, j \leqslant N - m.$$

The state ε_0 is persistent, but the others are transient.

5. In the preceding problems, let $N = 8$, $m = 4$, and suppose there are initially 5 white balls in the urn. What is the probability that no white balls are left after 2 drawings (of 4 balls each)?

6. A particle moves randomly along the interval $[1, m]$, coming to rest only at the points with coordinates $x = 1, \ldots, m$. The particle's motion is described by a Markov chain such that

$$p_{12} = 1, \qquad p_{m,m-1} = 1,$$
$$p_{j,j+1} = p, \qquad p_{j,j-1} = q \qquad (j = 2, \ldots, m-1),$$

with all other transition probabilities equal to zero. Which states are persistent and which transient?

7. In the preceding problem, show that the limiting probabilities defined in Theorem 7.4 do not exist. In particular, show that the condition (7.20) does not hold for any N.

Hint. $p_{11}(n) = 0$ if n is odd, while $p_{12}(n) = 0$ if n is even.

8. Consider the same kind of random walk as in Problem 6, but now suppose the nonzero transition probabilities are

$$p_{11} = q, \qquad p_{mm} = p,$$
$$p_{j,j+1} = p, \qquad p_{j,j-1} = q \qquad (j = 1, \ldots, m),$$

permitting the particle to stay at the points $x = 1$ and $x = m$. Which states are persistent and which transient? Show that the limiting probabilities $p_1^*, \ldots, p_m^*$ defined in Theorem 7.4 now exist.

9. In the preceding problem, calculate the limiting probabilities $p_1^*, \ldots, p_m^*$.

Ans. Solving the system of equations

$$p_1^* = qp_1^* + qp_2^*,$$
$$p_j^* = pp_{j-1}^* + qp_{j+1}^* \qquad (j = 2, \ldots, m-1),$$
$$p_m^* = pp_{m-1}^* + pp_m^*,$$

we get

$$p_j^* = \left(\frac{p}{q}\right)^{j-1} p_1^* \qquad (j = 1, \ldots, m).$$

Therefore

$$p_j^* = \frac{1}{m}$$

if $p = q$, while

$$p_j^* = \frac{1 - (p/q)}{1 - (p/q)^m} \left(\frac{p}{q}\right)^{j-1}$$

if $p \neq q$ (impose the condition that $\sum_{j=1}^{m} p_j^* = 1$).

10. Two marksmen A and B take turns shooting at a target. It is agreed that A will shoot after each hit, while B will shoot after each miss. Suppose A hits the target with probability $\alpha > 0$, while B hits the target with probability $\beta > 0$, and let n be the number of shots fired. What is the limiting probability of hitting the target as $n = \infty$?

Ans. $\dfrac{\beta}{1 - \alpha - \beta}$.

11. Suppose the condition (7.20) holds for a transition probability matrix whose column sums (as well as row sums) all equal unity. Find the limiting probabilities $p_1^*, \ldots, p_m^*$.

Ans. $p_1^* = \cdots = p_m^* = \dfrac{1}{m}$.

12. Suppose m white balls and m black balls are mixed together and divided equally between two urns. A ball is then drawn at random from each urn and put into the other urn. Suppose this is done n times. If the number of white balls in a given urn is j, we say that the "system" is in the state ε_j (the number of white balls in the other urn is then $m - j$). Prove that the limiting probabilities $p_0^*, p_1^*, \ldots, p_m^*$ defined in Theorem 7.4 exist, and calculate them.

Hint. The only nonzero transition probabilities are

$$p_{jj} = \frac{2j(m-j)}{m^2}, \qquad p_{j,j+1} = \frac{(m-j)^2}{m^2}, \qquad p_{j,j-1} = \frac{j^2}{m^2}.$$

Ans. Solving the system

$$p_j^* = p_{j-1}^* p_{j-1,j} + p_j^* p_{jj} + p_{j+1}^* p_{j+1,j} \qquad (j = 0, 1, \ldots, m),$$

we get $p_j^* = (C_j^m)^2 p_0^*$, and hence

$$p_j^* = \frac{(C_j^m)^2}{\sum_{j=0}^{m} (C_j^m)^2} = \frac{(C_j^m)^2}{C_m^{2m}} \qquad (j = 0, 1, \ldots, m)$$

(recall Problem 17, p. 12).

13. Find the stationary distribution $p_0^0, p_1^0, \ldots$ for the Markov chain whose only nonzero transition probabilities are

$$p_{j1} = \frac{j}{j+1}, \qquad p_{j,j+1} = \frac{1}{j+1} \qquad (j = 1, 2, \ldots).$$

Ans. $p_j^0 = \dfrac{1}{(e-1)j!}$.

14. Two gamblers A and B repeatedly play a game such that A's probability of winning is p, while B's probability of winning is $q = 1 - p$. Each bet is a dollar, and the total capital of both players is m dollars. Find the probability of each player being ruined, given that A's initial capital is j dollars.

Hint. Let ε_j denote the state in which A has j dollars. Then the situation is described by a Markov chain whose only nonzero transition probabilities are

$$p_{00} = 1, \qquad p_{mm} = 1,$$
$$p_{j,j+1} = p, \qquad p_{j,j-1} = q \qquad (j = 1, \ldots, m-1).$$

Ans. Let $\hat{p}_j = \lim_{n\to\infty} p_{j0}(n)$ be the probability of A's ruin, starting with an initial capital of j dollars. Then

$$\hat{p}_1 = p\hat{p}_2 + q, \qquad \hat{p}_{m-1} = q\hat{p}_{m-2},$$
$$\hat{p}_j = q\hat{p}_{j-1} + p\hat{p}_{j+1} \qquad (j = 2, \ldots, m-2)$$

(why?). Solving this system of equations, we get

$$\hat{p}_j = 1 - \frac{j}{m} \tag{7.33}$$

if $p = q$ (as in Example 3, p. 29), and

$$\hat{p}_j = \frac{1 - (p/q)^{m-j}}{1 - (p/q)^m} \tag{7.34}$$

if $p \neq q$. The probability of B's ruin is $1 - \hat{p}_j$.

15. In the preceding problem, prove that if $p > q$, then A's probability of ruin increases if the stakes are doubled.

16. Prove that a gambler playing against an adversary with unlimited capital is certain to be ruined unless his probability of winning in each play of the game exceeds $\frac{1}{2}$.

Hint. Let $m \to \infty$ in (7.33) and (7.34).

8

CONTINUOUS MARKOV PROCESSES

18. Definitions. The Sojourn Time

Consider a physical system with the following properties, which are the exact analogues of those given on p. 83 for a Markov chain, except that now the time t varies continuously:

a) The system can occupy any of a finite or countably infinite number of states $\varepsilon_1, \varepsilon_2, \ldots$
b) Starting from some initial state at time $t = 0$, the system changes its state randomly at subsequent times. Thus, the evolution of the system in time is described by the "random function" $\xi(t)$, equal to the state of the system at time t.[1]
c) At time $t = 0$, the system occupies the state ε_i with *initial probability*

$$p_i^0 = \mathbf{P}\,\{\xi(0) = \varepsilon_i\}, \qquad i = 1, 2, \ldots$$

d) Suppose the system is in the state ε_i at any time s. Then the probability that the system goes into the state ε_j after a time t is given by

$$p_{ij}(t) = \mathbf{P}\,\{\xi(s + t) = \varepsilon_j \mid \xi(s) = \varepsilon_i\}, \qquad i, j = 1, 2, \ldots, \tag{8.1}$$

regardless of its behavior before the time s. The numbers $p_{ij}(t)$, called the *transition probabilities*, do not depend on the time s.

A random process described by this model is called a *continuous Markov*

[1] Recall footnote 1, p. 83. Note that $\xi(t)$ is a random variable for any fixed t.

process[2] or simply a *Markov process* (as opposed to a Markov chain, which might be called a "discrete Markov process").

Let

$$p_j(t) = \mathbf{P}\{\xi(t) = \varepsilon_j\}, \qquad j = 1, 2, \ldots$$

be the probability that the system will be in the state ε_j at time t. Then, by arguments which hardly differ from those given on p. 84, we have

$$p_j(0) = p_j^0, \qquad j = 1, 2, \ldots,$$

$$p_j(s + t) = \sum_k p_k(s)p_{kj}(t), \qquad j = 1, 2, \ldots \tag{8.2}$$

and

$$p_{ij}(0) = \begin{cases} 1 & \text{if } j = i, \\ 0 & \text{if } j \neq i, \end{cases} \tag{8.3}$$

$$p_{ij}(s + t) = \sum_k p_{ik}(s)p_{kj}(t), \qquad i, j = 1, 2, \ldots \tag{8.4}$$

for arbitrary s and t [cf. (7.5) and (7.7)].

THEOREM 8.1. *Given a Markov process in the state ε at time $t = t_0$, let τ be the (random) time it takes the process to leave ε by going to some other state.*[3] *Then*

$$\mathbf{P}\{\tau > t\} = e^{-\lambda t}, \qquad t \geqslant 0, \tag{8.5}$$

where λ is a nonnegative constant.

Proof. Clearly $\mathbf{P}\{\tau > t\}$ is some function of t, say

$$\varphi(t) = \mathbf{P}\{\tau > t\}, \qquad t \geqslant 0.$$

If $\tau > s$, then the process will be in the same state at time $t_0 + s$ as at time t_0, and hence its subsequent behavior will be the same as if $s = 0$. In particular,

$$\mathbf{P}\{\tau > s + t \mid \tau > s\} = \varphi(t)$$

is the probability of the event $\{\tau > s + t\}$ given that $\tau > s$. It follows that

$$\mathbf{P}\{\tau > s + t\} = \mathbf{P}\{\tau > s + t \mid \tau > s\}\mathbf{P}\{\tau > s\} = \varphi(t)\varphi(s),$$

and hence

$$\varphi(s + t) = \varphi(s)\varphi(t)$$

or equivalently

$$\ln \varphi(s + t) = \ln \varphi(s) + \ln \varphi(t)$$

---

[2] More exactly, a *continuous Markov process with stationary transition probabilities*. where we allude to the fact that the numbers (8.1) do not depend on s (cf. footnote 2, p. 84).

[3] Here we prefer to talk about states of the *process* rather than states of the *system* (as in Chap. 7).

for arbitrary s and t. Therefore $\ln \varphi(t)$ is proportional to t (recall footnote 4, p. 40), say

$$\ln \varphi(t) = -\lambda t, \quad t \geqslant 0, \tag{8.6}$$

where λ is some nonnegative constant (why nonnegative?). But (8.6) implies (8.5). ∎

The parameter λ figuring in (8.4) is called the *density* of the transition out of the state ε. If $\lambda = 0$, the process remains forever in ε. If $\lambda > 0$, the probability of the process undergoing a change of state in a small time internal Δt is clearly

$$1 - \varphi(\Delta t) = \lambda\, \Delta t + o(\Delta t), \tag{8.7}$$

where $o(\Delta t)$ denotes an infinitesimal of higher order than Δt.

It follows from (8.5) that

$$\mathbf{P}\,\{t_1 < t \leqslant \tau_2\} = \varphi(t_1) - \varphi(t_2) = e^{-\lambda t_1} - e^{-\lambda t_2} = \int_{t_1}^{t_2} \lambda e^{-\lambda t}\, dt \tag{8.8}$$

for arbitrary nonnegative t_1 and t_2 ($t_1 \leqslant t_2$). Therefore the random variable τ, called the *sojourn time in state* ε, has the probability density

$$p_\tau(t) = \begin{cases} \lambda e^{-\lambda t} & \text{if } \ t \geqslant 0, \\ 0 & \text{if } \ t < 0. \end{cases} \tag{8.9}$$

The distribution corresponding to (8.8) and (8.9) is called the *exponential distribution*, with parameter λ. The mean value $\mathbf{E}\tau$, i.e., the "expected sojourn time in state ε," is given by

$$\mathbf{E}\tau = \int_0^\infty t p_\tau(t)\, dt = \frac{1}{\lambda}\,.$$

***Example* (*Radioactive decay*).** In Example 3, p. 58, we gave a probabilistic model of the radioactive decay of radium (Ra) into radon (Rn). The behavior of each of the n_0 radium atoms is described by a Markov process with two states (Ra and Rn) and one possible transition (Ra → Rn). As on p. 58, let $p(t)$ be the probability that a radium atom decays into a radon atom in time t, and $\xi(t)$ the number of alpha particles emitted in t seconds. Then, according to formula (5.7),

$$\mathbf{P}\,\{\xi(t) = k\} = \frac{a^k}{k!}\, e^{-a}, \qquad k = 0, 1, 2, \ldots,$$

where

$$a = \mathbf{E}\xi(t) = n_0 p(t).$$

It follows from (8.5) that

$$p(t) = 1 - e^{-\lambda t}, \qquad t \geqslant 0,$$

where λ is the density of the transition Ra $\to$ Rn. Recalling (8.7), we see that λ is the constant such that the probability of the transition Ra $\to$ Rn in a small time interval Δt equals $\lambda\Delta t + o(\Delta t)$.

The number of (undisintegrated) radium atoms left after time t is clearly $n_0 - \xi(t)$, with mean value

$$n(t) = \mathbf{E}[n_0 - \xi(t)] = n_0 - n_0 p(t) = n_0 e^{-\lambda t}, \qquad t \geqslant 0. \tag{8.10}$$

Let T be the *half-life* of radium, i.e., the amount of time required for half the radium to disappear (on the average). Then

$$n(T) = \frac{1}{2} n_0, \tag{8.11}$$

and hence, comparing (8.10) and (8.11), we find that T is related to the density λ of the transition Ra $\to$ Rn by the formula

$$T = \frac{\ln 2}{\lambda}.$$

19. The Kolmogorov Equations

Next we find differential equations satisfied by the transition probabilities of a Markov process:

THEOREM 8.2. *Given a Markov process with a finite number of states, suppose the transition probabilities $p_{ij}(t)$ are such that*[4]

$$\begin{aligned} 1 - p_{ii}(\Delta t) &= \lambda_i \, \Delta t + o(\Delta t), \qquad i = 1, 2, \ldots, \\ p_{ij}(\Delta t) &= \lambda_{ij} \, \Delta t + o(\Delta t), \qquad j \neq i, \quad i, j = 1, 2, \ldots, \end{aligned} \tag{8.12}$$

and let

$$\lambda_{ii} = -\lambda_i, \qquad i = 1, 2, \ldots \tag{8.13}$$

Then the transition probabilities satisfy two systems of linear differential equations, for ***forward Kolmogorov equations***[5]

$$p'_{ij}(t) = \sum_k p_{ik}(t)\lambda_{kj}, \qquad i, j = 1, 2, \ldots \tag{8.14}$$

and the ***backward Kolmogorov equations***

$$p'_{ij}(t) = \sum_k \lambda_{ik} p_{kj}(t), \qquad i, j = 1, 2, \ldots, \tag{8.15}$$

subject to the initial conditions (8.3).

---

[4] We might call λ_i the "density of the transition out of the state ε_i," and λ_{ij} the "density of the transition from the state ε_i to the state ε_j."

[5] The prime denotes differentiation with respect to t.

Proof. It follows from (8.4) that

$$p_{ij}(t+\Delta t) = \sum_k p_{ik}(t)p_{kj}(\Delta t) = \sum_k p_{ik}(\Delta t)p_{kj}(t).$$

Hence, using (8.12) and (8.13), we have

$$\frac{p_{ij}(t+\Delta t) - p_{ij}(t)}{\Delta t} = \sum_k p_{ik}(t)\left[\lambda_{kj} + \frac{o(\Delta t)}{\Delta t}\right] = \sum_k \left[\lambda_{ik} + \frac{o(\Delta t)}{\Delta t}\right]p_{kj}(t).$$

Both sums have definite limits as $\Delta t \to 0$. In fact,

$$\lim_{\Delta t\to 0} \sum_k p_{ik}(t)\left[\lambda_{kj} + \frac{o(\Delta t)}{\Delta t}\right] = \sum_k p_{ik}(t)\lambda_{kj}, \tag{8.16}$$

$$\lim_{\Delta t\to 0} \sum_k \left[\lambda_{ik} + \frac{o(\Delta t)}{\Delta t}\right]p_{kj}(t) = \sum_k \lambda_{ik}p_{kj}(t). \tag{8.17}$$

Therefore

$$\lim_{\Delta t\to 0} \frac{p_{ij}(t+\Delta t) - p_{ij}(t)}{\Delta t} = p_{ij}'(t)$$

also exists, and equals (8.16) and (8.17). ∎

Remark 1. It follows from (8.12) and the condition

$$\sum_j p_{ij}(\Delta t) = 1$$

that

$$\sum_{j\neq i} \lambda_{ij} = \lambda_i. \tag{8.18}$$

Remark 2. The Kolmogorov equations hold not only in the case of a finite number of states, but also in the case of a countably infinite number of states $\varepsilon_1, \varepsilon_2, \ldots$ if we make certain additional assumptions. In fact, suppose the error terms $o(\Delta t)$ in (8.12) are such that

$$\frac{o(\Delta t)}{\Delta t} \to 0 \quad \text{as} \quad \Delta t \to 0$$

uniformly in all i and j. Then the forward equations (8.14) hold if for any fixed j, there is a constant $C < \infty$ such that

$$\lambda_{ij} < C, \qquad i = 1, 2, \ldots,$$

while the backward equations (8.15) hold if the series (8.18) converges.

***Example* 1 (*The Poisson process*).** As in Example 4, p. 73, consider a "random flow of events" with density λ, and let $\xi(t)$ be the number of events which occur in time t. Then $\xi(t)$ is called a *Poisson process*. Clearly $\xi(t)$ is a Markov process, whose states can be described by the integers 0, 1, 2, . . . Moreover, $\xi(t)$ can only leave the state i by going into the state $i + 1$.

Therefore the transition densities λ_{ij} are just

$$\lambda_{ij} = \begin{cases} \lambda & \text{if } j = i + 1, \\ 0 & \text{if } j \neq i, i + 1, \end{cases}$$

$$\lambda_{ii} = -\lambda,$$

where we use (8.13) and (8.18).

The transition probabilities $p_{ij}(t)$ of the Poisson process $\xi(t)$ clearly satisfy the condition

$$p_{ij}(t) = p_{0,j-i}(t)$$

(why?). Let

$$p_j(t) = p_{0j}(t), \qquad j = 0, 1, 2, \ldots$$

Then the forward Kolmogorov equations take the form

$$\begin{aligned} p_0'(t) &= -\lambda p_0(t), \\ p_j'(t) &= \lambda p_{j-1}(t) - \lambda p_j(t), \qquad j = 1, 2, \ldots \end{aligned}$$

Introducing the new functions

$$f_j(t) = e^{\lambda t} p_j(t), \qquad j = 0, 1, 2, \ldots,$$

we find that

$$\begin{aligned} f_0'(t) &= \lambda f_0(t) + e^{\lambda t} p_0'(t) = \lambda f_0(t) - \lambda e^{\lambda t} p_0(t) = 0, \\ f_j'(t) &= \lambda f_j(t) + e^{\lambda t} p_j'(t) \\ &= \lambda f_j(t) + \lambda e^{\lambda t} p_{j-1}(t) - \lambda e^{\lambda t} p_j(t) = \lambda f_{j-1}(t), \qquad j = 1, 2, \ldots, \end{aligned}$$

where

$$\begin{aligned} f_0(0) &= 1, \\ f_j(0) &= 0, \qquad j = 1, 2, \ldots, \end{aligned} \tag{8.19}$$

because of (8.3). But the solution of the system of differential equations

$$\begin{aligned} f_0'(t) &= 0, \\ f_j'(t) &= \lambda f_{j-1}(t), \qquad j = 1, 2, \ldots, \end{aligned}$$

subject to the initial conditions (8.19), is obviously

$$f_0(t) = 1, \quad f_1(t) = \lambda t, \ldots, \quad f_n(t) = \frac{(\lambda t)^n}{n!}, \ldots$$

Returning to the original functions $p_j(t) = e^{-\lambda t} f_j(t)$, we find that

$$p_j(t) = \frac{(\lambda t)^j}{j!} e^{-\lambda t}, \qquad j = 0, 1, 2, \ldots$$

or equivalently

$$\mathbf{P}\{\xi(t) = j\} = \frac{(\lambda t)^j}{j!} e^{-\lambda t}, \qquad j = 0, 1, 2, \ldots,$$

just as on p. 75.

***Example* 2 (*A service system with exponential holding times*).** Consider a random flow of service calls arriving at a server, where the incoming "traffic" is of the Poisson type described in Example 1, with density λ. Thus $\lambda \Delta t + o(\Delta t)$ is the probability that at least one call arrives in a small time interval Δt. Suppose it takes a random time τ to service each incoming call, where τ has an exponential distribution with parameter μ:

$$\mathbf{P}\{\tau > t\} = e^{-\mu t} \tag{8.20}$$

(the case of "exponential holding times"). Then the service system has two states, a state ε_0 if the server is "free" and a state ε_1 if the server is "busy." It will be assumed that a call is rejected (and is no longer a candidate for service) if it arrives when the server is busy.

Suppose the system is in the state ε_0 at time t_0. Then its subsequent behavior does not depend on its previous history, since the calls arrive independently. The probability $p_{01}(\Delta t)$ of the system going from the state ε_0 to the state ε_1 during a small time interval Δt is just the probability $\lambda \Delta t + o(\Delta t)$ of at least one call arriving during Δt. Hence the density of the transition from ε_0 to ε_1 equals λ. On the other hand, suppose the system is in the state ε_1 at time t_1. Then the probability $p_{10}(t)$ of the system going from the state ε_1 to the state ε_0 after a time t is just the probability that service will fail to last another t seconds.[6] Suppose that at the time t_1, service has already been in progress for exactly s seconds. Then

$$p_{10}(t) = 1 - \mathbf{P}\{\tau > s + t \mid \tau > s\} = 1 - \frac{\mathbf{P}\{\tau > s + t\}}{\mathbf{P}\{\tau > s\}}.$$

Using (8.20), we find that

$$p_{10}(t) = 1 - \frac{e^{-\mu(s+t)}}{e^{-\mu s}} = 1 - e^{-\mu t}, \tag{8.21}$$

regardless of the time s, i.e., regardless of the system's behavior before the time t_1.[7] Hence the system can be described by a Markov process, with two states ε_0 and ε_1.

The transition probabilities of this Markov process obviously satisfy the conditions

$$p_{01}(t) = 1 - p_{00}(t), \qquad p_{10}(t) = 1 - p_{11}(t). \tag{8.22}$$

Moreover,

$$\lambda_{00} = -\lambda, \qquad \lambda_{01} = \lambda,$$

$$\lambda_{10} = \mu, \qquad \lambda_{11} = -\mu,$$

---

[6] For simplicity, we choose seconds as the time units.

[7] It is important to note that this is true only for exponential holding times (see W. Feller, *op. cit.*, p. 458).

where we use the fact that

$$p_{10}(\Delta t) = 1 - e^{-\mu \Delta t} = \mu \, \Delta t + o(\Delta t).$$

Hence in this case the forward Kolmogorov equations (8.14) become[8]

$$p_{00}'(t) = \lambda_{00} p_{00}(t) + \lambda_{10} p_{01}(t) = -\lambda p_{00}(t) + \mu[1 - p_{00}(t)],$$
$$p_{11}'(t) = \lambda_{01} p_{10}(t) + \lambda_{11} p_{11}(t) = \lambda[1 - p_{00}(t)] - \mu p_{11}(t),$$

i.e.,

$$\begin{aligned} p_{00}'(t) + (\lambda + \mu) p_{00}(t) &= \mu, \\ p_{11}'(t) + (\lambda + \mu) p_{11}(t) &= \lambda. \end{aligned} \tag{8.23}$$

Solving (8.23) subject to the initial conditions

$$p_{00}(0) = p_{11}(0) = 1,$$

we get

$$\begin{aligned} p_{00}(t) &= \left(1 - \frac{\mu}{\lambda + \mu}\right) e^{-(\lambda+\mu)t} + \frac{\mu}{\lambda + \mu}, \\ p_{11}(t) &= \left(1 - \frac{\lambda}{\lambda + \mu}\right) e^{-(\lambda+\mu)t} + \frac{\lambda}{\lambda + \mu}. \end{aligned} \tag{8.24}$$

20. More on Limiting Probabilities. Erlang's Formula

We now prove the continuous analogue of Theorem 7.4:

THEOREM 8.3. *Let $\xi(t)$ be a Markov process with a finite number of states, $\varepsilon_1, \ldots, \varepsilon_m$, each accessible from every other state. Then*

$$\lim_{t \to \infty} p_j(t) = p_j^*,$$

where $p_j(t)$ is the probability of $\xi(t)$ being in the state ε_j at time t. The numbers p_j^*, $j = 1, \ldots, m$, called the ***limiting probabilities****, do not depend on the initial probability distribution and satisfy the inequalities*

$$\max_i |p_{ij}(t) - p_j^*| \leq Ce^{-Dt}, \qquad |p_j(t) - p_j^*| \leq Ce^{-Dt} \tag{8.25}$$

for suitable positive constants C and D.

Proof. The proof is virtually the same as that of Theorem 7.4 for Markov chains, once we verify that the continuous analogue of the condition (7.20), p. 93 is automatically satisfied. In fact, we now have

$$\min_{i,j} p_{ij}(t) = \delta(t) > 0 \tag{8.26}$$

[8] Because of (8.22), there is no need to write equations for $p_{01}(t)$ and $p_{10}'(t)$.

for all $t > 0$. To show this, we first observe that $p_{ii}(t)$ is positive for sufficiently small t, being a continuous function (why?) satisfying the condition $p_{ii}(0) = 1$. But, because of (8.4),

$$p_{ii}(s + t) \geqslant p_{ii}(s)p_{ii}(t)$$

for arbitrary s and t, and hence $p_{ii}(t)$ is positive for all t.

To show that $p_{ij}(t)$, $i \neq j$ is also positive for all t, thereby proving (8.26) and the theorem, we note that

$$p_{ij}(s) > 0$$

for some s, since ε_j is accessible from ε_i. But

$$p_{ij}(t) \geqslant p_{ij}(u)p_{jj}(t - u), \qquad u \leqslant t,$$

again by (8.4), where, as just shown, $p_{jj}(t - u)$ is always positive. Hence it suffices to show that $p_{ij}(u) > 0$ for some $u \leqslant t$. Consider a Markov chain with the same states $\varepsilon_1, \ldots, \varepsilon_m$ and transition probabilities

$$p_{ij} = p_{ij}\left(\frac{s}{n}\right),$$

where n is an integer such that

$$n \geqslant m\frac{s}{t}.$$

Since

$$p_{ij}\left(n\frac{s}{n}\right) > 0,$$

the state ε_j is accessible from ε_i. But it is easy to see that ε_j is accessible from ε_i not only in n steps, but also in a number of steps n_0 no greater than the total number of states m (think this through). Therefore

$$p_{ij}\left(n_0\frac{s}{n}\right) > 0,$$

where

$$n_0\frac{s}{n} = u \leqslant t. \quad \blacksquare$$

The limiting probabilities p_j^*, $j = 1, \ldots, m$ form a stationary distribution in the same sense as on p. 96. More exactly, if we choose the initial distribution

$$p_j^0 = p_j^*, \qquad j = 1, \ldots, m,$$

then

$$p_j(t) \equiv p_j^*, \qquad j = 1, \ldots, m,$$

i.e., the probability of the system being in the state ε_j remains unchanged for all $t \geqslant 0$. In fact, taking the limit as $s \to \infty$ in (8.2), we get

$$p_j^* = \sum_i p_i^* p_{ij}(t), \qquad j = 1, \ldots, m. \tag{8.27}$$

But the right-hand side is just $p_j(t)$, as we see by choosing $s = 0$ in (8.2). Suppose the transition probabilities satisfy the conditions (8.12). Then differentiating (8.27) and setting $t = 0$, we find that

$$\sum_i p_i^* \lambda_{ij} = 0, \qquad j = 1, \ldots, m, \tag{8.28}$$

where λ_{ij} is the density of the transition from the state ε_i to the state ε_j.

Example **(*A service system with m servers*).** Consider a service system which can handle up to m incoming calls at once, i.e., suppose there are m servers and an incoming call can be handled if at least one server is free. As in Example 2, p. 108, we assume that the incoming traffic is of the Poisson type with density λ, and that the time it takes each server to service a call is exponentially distributed with parameter μ (this is again a case of "exponential holding times"). Moreover, it will be assumed that a call is rejected (and is no longer a candidate for service) if it arrives when all m servers are busy, and that the "holding times" of the m servers are independent random variables.

If precisely j servers are busy, we say that the service system is in the state ε_j ($j = 0, 1, \ldots, m$). In particular, ε_0 means that the whole system is free and ε_m that the system is completely busy. For almost the same reasons as on p. 108, the evolution of the system in time from state to state is described by a Markov process. The only nonzero transition probabilities of this process are

$$\begin{aligned} &\lambda_{00} = -\lambda, \quad \lambda_{01} = \lambda, \quad \lambda_{mm} = -m\mu, \\ &\lambda_{j,j-1} = j\mu, \quad \lambda_{jj} = -(\lambda + j\mu), \quad \lambda_{j,j+1} = \lambda \qquad (j = 1, \ldots, m-1). \end{aligned} \tag{8.29}$$

In fact, suppose the system is in the state ε_j. Then a transition from ε_j to ε_{j+1} takes place if a single call arrives, which happens in a small time interval Δt with probability $\lambda\Delta t + o(\Delta t)$.[9] Moreover, the probability that none of the j busy servers becomes free in time Δt is just

$$[1 - \mu\Delta t + o(\Delta t)]^j,$$

since the holding times are independent, and hence the probability of at least one server becoming free in time Δt equals

$$1 - [1 - \mu\Delta t + o(\Delta t)]^j = j\mu\Delta t + o(\Delta t).$$

[9] For small Δt, this is also the probability of at least one call arriving in Δt.

But for small Δt, this is also the probability of a single server becoming free in time Δt, i.e., of a transition from ε_j to ε_{j-1}. The transitions to new states other than ε_{j-1} or ε_{j+1} have small probabilities of order $o(\Delta t)$. These considerations, together with (8.12) and the formula

$$\sum_j \lambda_{ij} = 0$$

implied by (8.12) and (8.13), lead at once to (8.29).

In the case $m = 1$, it is clear from the formulas (8.24) that the transition probabilities $p_{ij}(t)$ approach their limiting values "exponentially fast" as $t \to \infty$. It follows from the general formula (8.25) that the same is true in the case $m > 1$ (more than 1 server). To find these limiting probabilities p_j^*, we use (8.28) and (8.29), obtaining the following system of linear equations:

$$\begin{aligned} \lambda p_0^* &= \mu p_1^*, \\ (\lambda + j\mu)p_j^* &= \lambda p_{j-1}^* + (j+1)\mu p_{j+1}^* \qquad (j = 1, \ldots, m-1), \\ \lambda p_{m-1}^* &= m\mu p_m^*. \end{aligned}$$

Solving this system, we get

$$p_j^* = \frac{1}{j!}\left(\frac{\lambda}{\mu}\right)^j p_0^*, \qquad j = 0, 1, \ldots, m.$$

Using the "normalization condition"

$$\sum_{j=0}^m p_j^* = 1$$

to determine p_0^*, we finally obtain *Erlang's formula*

$$p_j^* = \frac{\dfrac{1}{j!}\left(\dfrac{\lambda}{\mu}\right)^j}{\displaystyle\sum_{j=0}^m \frac{1}{j!}\left(\frac{\lambda}{\mu}\right)^j}, \qquad j = 0, 1, \ldots, m \tag{8.30}$$

for the limiting probabilities.

PROBLEMS

1. Suppose each alpha particle emitted by a sample of radium has probability p of being recorded by a Geiger counter. What is the probability of exactly n particles being recorded in t seconds?

Ans. $\dfrac{(\lambda pt)^n}{n!} e^{-\lambda pt}$, where λ is the same as in the example on p. 104.

2. A man has two telephones on his desk, one receiving calls with density λ_1, the other with density λ_2.[10] What is the probability of exactly n calls being received in t seconds?

Hint. Recall Problem 9, p. 81. Neglect the effect of the lines being found busy.

Ans. $\dfrac{[(\lambda_1 + \lambda_2)t]^n}{n!} e^{-(\lambda_1+\lambda_2)t}.$

3. Given a Poisson process with density λ, let $\xi(t)$ be the number of events occurring in time t. Find the correlation coefficient of the random variables $\xi(t)$ and $\xi(t + \tau)$, where $\tau > 0$.

Ans. $\sqrt{\dfrac{t}{t + \tau}}.$

4. Show that (8.24) leads to Erlang's formula (8.30) for $m = 1$.

5. The arrival of customers at the complaint desk of a department store is described by a Poisson process with density λ. Suppose each clerk takes a random time τ to handle a complaint, where τ has an exponential distribution with parameter μ, and suppose a customer leaves whenever he finds all the clerks busy. How many clerks are needed to make the probability of customers leaving unserved less than 0.015 if $\lambda = \mu$?

Hint. Use Erlang's formula (8.30).

Ans. Four.

6. A single repairman services m automatic machines, which normally do not require his attention. Each machine has probability $\lambda\Delta t + o(\Delta t)$ of breaking down in a small time interval Δt. The time required to repair each machine is exponentially distributed with parameter μ. Find the limiting probability of exactly j machines being out of order.

Hint. Solve the system of equations

$$\begin{aligned} m\lambda p_0^* &= \mu p_1^*, \\ [(m - j)\lambda + \mu]p_j^* &= (m - j + 1)\lambda p_{j-1}^* + \mu p_{j+1}^*, \\ \mu p_m^* &= \lambda p_{m-1}^*. \end{aligned}$$

Ans. $p_j^* = \dfrac{m!}{(m - j)!}\left(\dfrac{\lambda}{\mu}\right)^j p_0^*, \qquad j = 0, 1, \ldots, m,$

where p_0^* is determined from the condition $\sum_{j=0}^{m} p_j^* = 1.$

Comment. Note the similarity between this result and formula (8.30).

7. In the preceding problem, find the average number of machines awaiting the repairman's attention.

Ans. $m - \dfrac{\lambda + \mu}{\lambda}(1 - p_0^*).$

---

[10] It is assumed that the incoming calls on each line form a Poisson process.

8. Solve Problem 6 for the case of r repairmen, where $1 < r < m$.

9. An electric power line serves m identical machines, each operating independently of the others. Suppose that in a small interval of time Δt each machine has probability $\lambda \Delta t + o(\Delta t)$ of being turned on and probability $\mu \Delta t + o(\Delta t)$ of being turned off. Find the limiting probability p_j^* of exactly j machines being on.

Hint. Solve the system of equations

$$\begin{aligned} m\lambda p_0^* &= \mu p_1^*, \\ [(m-j)\lambda + j\mu]p_j^* &= (m-j+1)\lambda p_{j-1}^* + (j+1)\mu p_{j+1}^*, \\ m\mu p_m^* &= \lambda p_{m-1}^*. \end{aligned}$$

Ans. $p_j^* = C_j^m \left(\dfrac{\mu}{\lambda + \mu}\right)^{m-j} \left(\dfrac{\lambda}{\lambda + \mu}\right)^j, \qquad j = 0, 1, \ldots, m.$

10. Show that the answer to the preceding problem is just what one would expect by an elementary argument if $\lambda = \mu$.

Appendix 1

INFORMATION THEORY

Given a random experiment with N equiprobable outcomes $A_1, \ldots, A_N$, how much "information" is conveyed on the average by a message $\mathscr{M}$ telling us which of the outcomes $A_1, \ldots, A_N$ has actually occurred? As a reasonable measure of this information, we might take the average length of the message $\mathscr{M}$, provided $\mathscr{M}$ is written in an "economical way." For example, suppose we use a "binary code," representing each of the possible outcomes $A_1, \ldots, A_N$ by a "code word" of length l, i.e., by a sequence

$$a_1 \ldots a_l,$$

where each "digit" a_k is either a 0 or a 1. Obviously there are 2^l such words (all of the same length l), and hence to be capable of uniquely designating the N possible outcomes, we must choose a value of l such that

$$N \leqslant 2^l. \tag{1}$$

The smallest value of l satisfying (1) is just the integer such that

$$0 \leqslant l - \log_2 N < 1.$$

This being the case, the quantity

$$I = \log_2 N \tag{2}$$

is clearly a reasonable definition of the average amount of information in the message $\mathscr{M}$ (measured in binary units or "bits").

More generally, suppose the outcomes $A_1, \ldots, A_N$ have different probabilities

$$p_1 = \mathbf{P}(A_1), \ldots, p_N = \mathbf{P}(A_N). \tag{3}$$

Then it is clear that being told about a rare outcome conveys more information than being told about a likely outcome.[1] To take this into account, we repeat the experiment n times, where n is very large, and send a new message $\mathscr{M}'$ conveying the result of the whole series of n trials. Each outcome is now a sequence

$$A_{i_1}, \ldots, A_{i_n}, \tag{4}$$

where A_{i_k} is the outcome occurring at the kth trial. Of the N^n possible outcomes of the whole series of trials, it is overwhelmingly likely that the outcome will belong to a much smaller set containing only

$$N_n = \frac{n!}{n_1! \cdots n_N!} \tag{5}$$

outcomes, where

$$n_1 = np_1, \ldots, n_N = np_N, \qquad n_1 + \cdots + n_N = n.$$

In fact, let $n_i = n(A_i)$ be the number of occurrences of the event A_i in N trials. Then

$$\frac{n_i}{n} \sim p_i$$

by the law of large numbers (see Sec. 12), and hence $n_i \sim np_i$. To get (5), we merely replace $\sim$ by $=$ and invoke Theorem 1.4, p. 7. We emphasize that this is a plausibility argument and not a rigorous proof,[2] but the basic idea is perfectly sound.

Continuing in this vein, we argue that only a negligibly small amount of information is lost on the average if we neglect all but the set of N_n highly likely outcomes of the form (4), all with the same probability

$$\mathbf{P}A(_{i_1}) \cdots \mathbf{P}(A_{i_n}) = p_1^{n_1} \cdots p_N^{n_N}.$$

This brings us back to the case of equiprobable outcomes, and suggests defining the average amount of information conveyed by the message $\mathscr{M}'$ as

$$I' = \log_2 N_n.$$

Hence, dividing by the number of trials, we find that the average amount of information in the original message $\mathscr{M}$ is just

$$I = \frac{\log_2 N_n}{n}. \tag{6}$$

[1] In particular, no information at all is conveyed by being told that the sure event has occurred, because we already know what the message will be!

[2] Among other missing details, we note that the numbers $n_1, \ldots, n_N$ are in general not all integers, as assumed in (4).

To calculate (6), we apply Stirling's formula (see p. 10) to the expression (5), obtaining

$$N_n \sim \frac{\sqrt{2\pi n}\, n^n e^{-n}}{\sqrt{2\pi n_1}\, n_1^{n_1} e^{-n_1} \cdots \sqrt{2\pi n_N}\, n_N^{n_N} e^{-n_N}},$$

and hence

$$\begin{aligned}\ln N_n &\sim n \ln n - np_1 \ln (np_1) - \cdots - np_N \ln (np_N)\\ &= n \ln n - (np_1 + \cdots + np_N) \ln n - np_1 \ln p_1 - \cdots - np_N \ln p_N\\ &= -n \sum_{i=1}^{N} p_i \ln p_i\end{aligned}$$

in terms of the natural logarithm

$$\ln x = \log_e x,$$

or equivalently

$$\log_2 N_n \sim -n \sum_{i=1}^{N} p_i \log_2 p_i \tag{7}$$

in terms of the logarithm to the base 2. Changing $\sim$ to $=$ and substituting (7) into (6), we get *Shannon's formula*

$$I = -\sum_{i=1}^{N} p_i \log_2 p_i \tag{8}$$

for the average amount of information in a message $\mathscr{M}$ telling which of the N outcomes $A_1, \ldots, A_N$ with probabilities (3) has occurred. Note that (8) reduces to (2) if the outcomes are equiprobable, since then

$$p_1 = \cdots = p_N = \frac{1}{N}.$$

***Example* 1 (*Average time of psychological reaction*).** One of N lamps is illuminated at random, where p_i is the probability of the ith lamp being turned on, and an observer is asked to point out the lamp which is lit. In a long series of independent trials it turns out[3] that the average time required to give the correct answer is proportional to the quantity (8) rather than to the number of lamps N, as might have been expected.

We can interpret the quantity (8) not only as the average amount of information conveyed by the message $\mathscr{M}$, but also the average amount of "uncertainty" residing in the given random experiment, and hence as a measure of the randomness of the experiment. Receiving the message reduces the uncertainty of the outcome of the experiment to zero, since the

[3] See A. M. Yaglom and I. M. Yaglom, *Wahrscheinlichkeit und Information*, second edition, VEB Deutscher Verlag der Wissenschaften, Berlin (1965), p. 67.

message tells us the result of the experiment with complete certainty. More generally, we might ask for the amount of information about one "full set" of mutually exclusive events $A_1, \ldots, A_N$ conveyed by being told which of a related full set of mutually exclusive events $B_1, \ldots, B_{N'}$ has occurred. Suppose the two sets of events have probabilities $\mathbf{P}(A_1), \ldots, \mathbf{P}(A_N)$ and $\mathbf{P}(B_1), \ldots, \mathbf{P}(B_{N'})$, where $\mathbf{P}(A_1) + \cdots + \mathbf{P}(A_N) = 1$,

$$\mathbf{P}(B_1) + \cdots + \mathbf{P}(B_{N'}) = 1.$$

Moreover, let $\mathbf{P}(A_iB_j)$ be the probability that both events A_i and B_j occur, while $\mathbf{P}(A_i \mid B_j)$ is the probability of A_i occurring if B_j is known to have occurred. Then

$$I_{A|B_j} = -\sum_{i=1}^{N} \mathbf{P}(A_i \mid B_j) \log_2 \mathbf{P}(A_i \mid B_j)$$

is the amount of uncertainty about the events $A_1, \ldots, A_N$ remaining after B_j is known to have occurred, and hence

$$\begin{aligned} I_{A|B} = -\sum_{j=1}^{N'} \mathbf{P}(B_j) I_{A|B_j} &= -\sum_{j=1}^{N'} \sum_{i=1}^{N} \mathbf{P}(B_j)\mathbf{P}(A_i \mid B_j) \log_2 \mathbf{P}(A_i \mid B_j) \\ &= -\sum_{i,j} \mathbf{P}(A_iB_j) \log_2 \frac{\mathbf{P}(A_iB_j)}{\mathbf{P}(B_j)} \end{aligned} \tag{9}$$

is the average amount of uncertainty about $A_1, \ldots, A_N$ remaining after it is known which of the events $B_1, \ldots, B_{N'}$ has occurred. Let I_{AB} be the information about the events $A_1, \ldots, A_N$ conveyed by knowledge of which of the events $B_1, \ldots, B_{N'}$ has occurred. Then clearly[4]

$$I_{AB} = I_A - I_{A|B}, \tag{10}$$

where

$$I_A = -\sum_{i=1}^{N} \mathbf{P}(A_i) \log_2 \mathbf{P}(A_i) = -\sum_{i=1}^{N} \sum_{j=1}^{N'} \mathbf{P}(A_iB_j) \log_2 \mathbf{P}(A_i) \tag{11}$$

is the quantity previously denoted by I (justify the last step). Combining (9) and (11), we finally get

$$I_{AB} = \sum_{i,j} \mathbf{P}(A_iB_j) \log_2 \frac{\mathbf{P}(A_iB_j)}{\mathbf{P}(A_i)P(B_j)}. \tag{12}$$

***Example* 2 (*Weather prediction*).** During a certain season it rains about once every five days, the weather being fair the rest of the time. Every night a prediction is made of the next day's weather. Suppose a prediction of rain is wrong about half the time, while a prediction of fair weather is wrong

---

[4] In words, (10) says that "the information in the message" equals "the uncertainty before the message is received" minus "the uncertainty after the message is received."

only about one time out of ten. How much information about the weather is conveyed on the average by the predictions?

Solution. Let A_1 denote rain, A_2 fair weather, B_1 a prediction of rain and B_2 a prediction of fair weather. Then, to a good approximation,

$$\mathbf{P}(A_1) = \frac{1}{5}, \qquad \mathbf{P}(A_2) = \frac{4}{5},$$

$$\mathbf{P}(A_1 \mid B_1) = \frac{1}{2}, \qquad \mathbf{P}(A_1 \mid B_2) = \frac{1}{10}.$$

Moreover, since

$$\mathbf{P}(A_1) = \mathbf{P}(A_1 \mid B_1)\mathbf{P}(B_1) + \mathbf{P}(A_1 \mid B_2)\mathbf{P}(B_2),$$

we have

$$\frac{1}{5} = \frac{1}{2}\mathbf{P}(B_1) + \frac{1}{10}[1 - \mathbf{P}(B_1)],$$

and hence

$$\mathbf{P}(B_1) = \frac{1}{4}, \qquad \mathbf{P}(B_2) = \frac{3}{4},$$

$$\mathbf{P}(A_1B_1) = \mathbf{P}(A_1 \mid B_1)\mathbf{P}(B_1) = \frac{1}{8},$$

$$\mathbf{P}(A_1B_2) = \mathbf{P}(A_1 \mid B_2)\mathbf{P}(B_2) = \frac{3}{40},$$

$$\mathbf{P}(A_2B_1) = [1 - \mathbf{P}(A_1 \mid B_1)]\mathbf{P}(B_1) = \frac{1}{8},$$

$$\mathbf{P}(A_2B_2) = [1 - \mathbf{P}(A_1 \mid B_2)]\mathbf{P}(B_2) = \frac{27}{40}.$$

It follows from (12) that

$$I_{AB} = \frac{1}{8}\log_2\frac{5}{2} + \frac{3}{40}\log_2\frac{1}{2} + \frac{1}{8}\log_2\frac{5}{8} + \frac{27}{40}\log_2\frac{9}{8} \approx 0.12$$

is the average amount of weather conveyed by a prediction. In the case of 100% accurate predictions, $A_1 = B_1$, $A_2 = B_2$ and (12) reduces to

$$I_{AB} = -\frac{1}{5}\log_2\frac{1}{5} - \frac{4}{5}\log_2\frac{4}{5} \approx 0.72.$$

PROBLEMS

1. Which conveys more information, a message telling a stranger's birthday or a message telling his telephone number?

2. Find the average amount of information in bits of a message telling whether or not the outcome of throwing a pair of unbiased dice is

a) An odd number; b) A prime number; c) A number no greater than 5.

3. An experiment has four possible outcomes, with probabilities $\frac{1}{16}$, $\frac{3}{16}$, $\frac{5}{16}$ and $\frac{7}{16}$, respectively. What is the average amount of uncertainty about the outcome of the experiment?

4. Each of the signals $A_1, \ldots, A_n$ has equal probability of being transmitted over a communication channel. In the absence of noise, the signal A_j is received as a_j ($j = 1, \ldots, n$), while in the presence of noise A_j has probability p of being received as a_j and equal probability of being received as any of the other symbols. What is the average amount of information about the symbols $A_1, \ldots, A_n$ conveyed by receiving one of the signals $a_1, \ldots, a_n$

a) In the absence of noise; b) In the presence of noise?

Ans. a) $\log_2 n$; b) $\log_2 n + p \log_2 p + (1 - p) \log_2 \frac{1 - p}{n - 1}$.

Appendix 2

GAME THEORY

Consider the following simple model of a game played repeatedly by two players.[1] Each player can choose one of two strategies determining the result of the game. The interests of the players are completely conflicting, e.g., whatever one player wins, the other loses.[2] Such a "two-person game" can be described by the table shown in Figure 9, where the quantity in the ith row and jth column is the amount gained by the first player if he chooses strategy i while his opponent chooses strategy j $(i, j = 1, 2)$. For example, v_{12} is the amount gained by the first player (the first player's "payoff") if he chooses the first strategy and his opponent (the second player) chooses the second strategy, while $-v_{21}$ is the second player's payoff if he chooses strategy 1 and his opponent (the first player) chooses strategy 2. It is now natural to ask for each player's "optimal strategy."

v_{11}	v_{12}
v_{21}	v_{22}

FIGURE 9

This question is easily answered in the case where

$$\min(v_{11}, v_{12}) \geqslant \max(v_{21}, v_{22}), \tag{1}$$

say, since then regardless of how the second player acts, the first player

[1] More generally, a game of strategy involves more than two players, each with more than two available strategies, but the essential features of game theory (in particular, its connection with probability) emerges even in this extremely simple case.

[2] Such a game, in which the algebraic sum of the players' winnings is zero, is called a *zero-sum game*.

should always choose the first strategy, thereby guaranteeing himself a gain of at least

$$\min (v_{11}, v_{12}).$$

Assuming a "clever" opponent, the second player should then choose the strategy which minimizes the first player's maximum gain, i.e., the strategy j such that

$$v_{1j} = \min (v_{11}, v_{12}).$$

The case just described is atypical. Usually, a relation like (1) does not hold, and each player should adopt a "mixed strategy," sometimes choosing one of the two "pure strategies" available to him and sometimes choosing the other, with definite probabilities (found in a way to be discussed). More exactly, the first player should choose the ith strategy with probability p_{1i}, while the second player should (independently) choose the jth strategy with probability p_{2j}. Then the first player's strategy is described by a probability distribution $P_1 = \{p_{11}, p_{12}\}$, while the second player's strategy is described by a probability distribution $P_2 = \{p_{21}, p_{22}\}$. If these mixed strategies are adopted, the average gain to the first player is clearly just

$$V(P_1, P_2) = \sum_{i,j=1}^{2} v_{ij} p_{1i} p_{2j}. \tag{2}$$

Suppose the second player makes the optimal response to each strategy $P_1 = \{p_{11}, p_{12}\}$ chosen by the first player, by adopting the strategy $P_2^* = \{p_{21}^*, p_{22}^*\}$ minimizing the first player's gain. The first player then wins an amount

$$V(P_1, P_2^*) = \min_{P_2} V(P_1, P_2) = V_1(P_1)$$

if he chooses the strategy P_1. To maximize this gain, the first player should choose the strategy $P_1^0 = \{p_{11}^0, p_{12}^0\}$ such that

$$V_1(P_1^0) = \max_{P_1} V_1(P_1),$$

always, of course, under the assumption that his opponent plays in the best possible way. Exactly the same argument can be applied to the second player, and shows that his optimal strategy, guaranteeing his maximum average gain under the assumption of optimal play on the part of his opponent, is the strategy $P_2^0 = \{p_{21}^0, p_{22}^0\}$ such that

$$V_2(P_2^0) = \max_{P_2} V_2(P_2),$$

where

$$V_2(P_2) = \min_{P_1} \{-V(P_1, P_2)\}.$$

To calculate the optimal strategies P_1^0 and P_2^0, we consider the function

$$V(x, y) = v_{11}xy + v_{12}x(1 - y) + v_{21}(1 - x)y + v_{22}(1 - x)(1 - y),$$

which for $x = p_{11}$ and $y = p_{21}$ equals the average gain of the first player if the mixed strategies $P_1 = \{p_{11}, p_{12}\}$ and $P_2 = \{p_{21}, p_{22}\}$ are chosen. The function $V(x, y)$ is linear in each of the variables x and y, $0 \leqslant x, y \leqslant 1$. Hence, for every fixed x, $V(x, y)$ achieves its minimum $V_1(x)$ at one of the end points of the interval $0 \leqslant y \leqslant 1$, i.e., for $y = 0$ or $y = 1$:

$$V_1(x) = \min_y V(x, y) = \min \{v_{12}x + v_{22}(1 - x), v_{11}x + v_{21}(1 - x)\}.$$

As shown in Figure 10, the graph of the function $V_1(x)$ is a broken line with

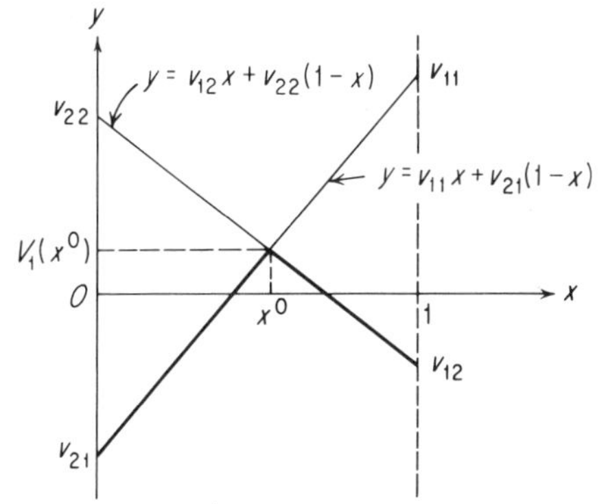

FIGURE 10. A case where $\min(v_{11}, v_{12}) < \max(v_{21}, v_{22})$.

vertex at the point x^0 such that

$$v_{12}x^0 + v_{22}(1 - x^0) = v_{11}x^0 + v_{21}(1 - x^0),$$

i.e., at the point

$$x^0 = \frac{v_{22} - v_{21}}{v_{11} + v_{22} - (v_{12} + v_{21})}. \tag{3}$$

The value $x = x^0$ for which the function $V_1(x)$, $0 \leqslant x \leqslant 1$ takes its maximum is just the probability p_{11}^0 with which the first player should choose his first pure strategy. The corresponding optimal mixed strategy $P_1^0 = \{p_{11}^0, p_{12}^0\}$ guarantees the maximum average gain for the first player under the assumption of optimal play on the part of his opponent. This gain is

$$V_1(x^0) = v_{11}x^0 + v_{21}(1 - x^0) = v_{12}x^0 + v_{22}(1 - x^0). \tag{4}$$

Moreover, (4) implies

$$V(x^0, y) = y[v_{11}x^0 + v_{21}(1 - x^0)] + (1 - y)[v_{12}x^0 + v_{22}(1 - x^0)] = V_1(x^0)$$

for any y in the interval $0 \leqslant y \leqslant 1$. Hence, by choosing $p_{11}^0 = x^0$, the first player guarantees himself an average gain $V_1(x^0)$ regardless of the value of y, i.e., regardless of how his opponent plays. However, if the first player deviates from this optimal strategy, by choosing $p_{11} = x \neq x^0$, then his opponent need only choose $p_{21} = y$ equal to 0 or 1 (as the case may be) to reduce the first player's average gain to just $V_1(x)$.

Applying the same considerations to the second player, we find that the second player's optimal strategy is such that $p_{21}^0 = y^0$, where

$$y^0 = \frac{v_{11} - v_{12}}{v_{11} + v_{22} - (v_{12} + v_{21})} \tag{5}$$

[(5) is obtained from (3) by reversing the roles of players 1 and 2, i.e., by interchanging the indices 1 and 2]. As in the case of the first player, this choice guarantees the second player an average gain $V_2(y^0)$ regardless of the first player's strategy, i.e.,

$$-V(x, y) = V_2(y^0), \qquad 0 \leqslant x \leqslant 1. \tag{6}$$

In particular, (4) and (6) imply

$$V_1(x^0) = V(x^0, y^0),$$
$$V_2(y^0) = -V(x^0, y^0).$$

***Example* 1.** One player repeatedly hides either a dime or a quarter, and the second player guesses which coin is hidden. If he guesses properly, he gets the coin, but otherwise he must pay the first player 15 cents. Find both players' optimal strategies.

Solution. Here

$$v_{11} = -10, \qquad v_{12} = 15,$$
$$v_{21} = 15, \qquad v_{22} = -25,$$

so that, by (3),

$$p_{11}^0 = x^0 = \frac{-25 - 15}{-35 - 30} = \frac{8}{13}.$$

Therefore the first player should hide the dime with probability $\frac{8}{13}$, and hide the quarter with probability $\frac{5}{13}$.[3] Similarly, by (5),

$$p_{21}^0 = y^0 = \frac{-10 - 15}{-35 - 30} = \frac{5}{13},$$

[3] For the first player, hiding the dime is (pure) strategy 1, and hiding the quarter is strategy 2. For the second player, guessing the dime is strategy 1, and guessing the quarter is strategy 2.

and hence the second player should guess that the hidden coin is a dime with probability $\frac{5}{13}$, and that it is a quarter with probability $\frac{8}{13}$. Then, according to (4), the first player's average gain will be

$$V(x^0, y^0) = -10 \cdot \frac{8}{13} + 15 \cdot \frac{5}{13} = -\frac{5}{13},$$

while the second player's average gain will be

$$-V(x^0, y^0) = \frac{5}{13}.$$

Thus this game is unfavorable to the first player, who loses an average of $\frac{5}{13}$ cents every time he plays, even if he adopts the optimal strategy. However, any departure from the optimal strategy will lead to an even greater loss, if his opponent responds properly.

***Example* 2 (*Aerial warfare*).**[4] White repeatedly sends two-plane missions to attack one of Blue's installations. One plane carries bombs, and the other (identical in appearance) flies cover for the plane carrying the bombs. Suppose the lead plane can be defended better by the guns of the plane in the second position than vice versa, so that the chance of the lead plane surviving an attack by Blue's fighter is 80%, while the chance of the plane in the second position surviving such an attack is only 60%. Suppose further that Blue can attack just one of White's planes and that Blue's sole concern is the protection of his installation, while White's sole concern is the destruction of Blue's installation. Which of White's planes should carry the bombs, and which plane should Blue attack?

Solution. Let White's payoff be the probability of accomplishing the mission. Then[5]

$$v_{11} = 0.8, \qquad v_{12} = 1,$$
$$v_{21} = 1, \qquad v_{22} = 0.6,$$

and hence

$$p_{11}^0 = x^0 = \frac{-0.4}{-0.6} = \frac{2}{3}, \qquad p_{21}^0 = y^0 = \frac{-0.2}{-0.6} = \frac{1}{3},$$

by (3) and (5). Thus always putting the bombs in the lead plane is not White's best strategy, although this plane is less likely to be shot down than

[4] After J. D. Williams, *The Compleat Strategyst*, McGraw-Hill Book Co., Inc., New York (1954), p. 47.

[5] For White, putting the bombs in the lead plane is (pure) strategy 1, and putting the bombs in the other plane is strategy 2. For Blue, attacking the lead plane is strategy 1, and attacking the other plane is strategy 2.

the other. In fact, if White always puts the bombs in the lead plane, then Blue will always attack this plane and the resulting probability of the mission succeeding will be 0.8. On the other hand, if White adopts the optimal mixed strategy and puts the bombs in the lead plane only two times out of three, he will increase his probability of accomplishing the mission by $\frac{1}{15}$, since, according to (4),

$$V(x^0, y^0) = \frac{2}{3} \cdot \frac{8}{10} + \frac{1}{3} \cdot 1 = \frac{13}{15}.$$

By the same token, Blue's best strategy is to attack the lead plane only one time out of three and the other plane the rest of the time.

PROBLEMS

1. Prove that the game considered in Example 1 becomes favorable to the first player if the second player's penalty for incorrect guessing is raised to 20 cents.

2. In Example 1, let a be the second player's penalty for incorrect guessing. For what value of a does the game become "fair"?

3. Blue has two installations, only one of which he can successfully defend, while White can attack either but not both of Blue's installations. Find the optimal strategies for White and Blue if one of the installations is three times as valuable as the other.[6]

Ans. White should attack the less valuable installation 3 out of 4 times, while Blue should defend the more valuable installation 3 out of 4 times.

---

[6] After J. D. Williams, *op. cit.*, p. 51.

Appendix 3

BRANCHING PROCESSES

Consider a group of particles, each "randomly producing" more particles of the same type by the following process:

a) The probability that each of the particles originally present at some time $t = 0$ produces a group of k particles after a time t is given by $p_k(t)$, where $k = 0, 1, 2, \ldots$ and $p_k(t)$ is the same for all the particles.[1]
b) The behavior of each particle is independent of the behavior of the other particles and of the events prior to the initial time $t = 0$.

A random process described by this model is called a *branching process.* As concrete examples of such processes, think of nuclear chain reactions, survival of family names, etc.[2]

Let $\xi(t)$ be the total number of particles present at time t. Then $\xi(t)$ is a Markov process (why?). Suppose there are exactly k particles initially present at time $t = 0$, and let $\xi_i(t)$ be the number of particles produced by the ith particle after a time t. Then clearly

$$\xi(t) = \xi_1(t) + \cdots + \xi_k(t), \tag{1}$$

where the random variables $\xi_1(t), \ldots, \xi_k(t)$ are independent and have the same probability distribution

$$\mathbf{P}\,\{\xi_1(t) = n\} = p_n(t), \qquad n = 0, 1, 2, \ldots$$

[1] The case $k = 0$ corresponds to "annihilation" of a particle.
[2] Concerning these examples and others, see W. Feller, *op. cit.*, p. 294.

Let $p_{kn}(t)$ be the probability of the k particles giving rise to a total of n particles after time t, so that the numbers $p_{kn}(t)$ are the transition probabilities of the Markov process $\xi(t)$, and introduce the generating functions[3]

$$F(t, z) = \sum_{n=0}^{\infty} p_n(t) z^n, \tag{2}$$

$$F_k(t, z) = \sum_{n=0}^{\infty} p_{kn}(t) z^n. \tag{3}$$

Suppose the probability of a single particle giving rise to a total of n particles in a small time interval Δt is

$$p_n(\Delta t) = \lambda_n \Delta t + o(\Delta t),$$

while the probability of the particle remaining unchanged is

$$p_1(\Delta t) = 1 - \lambda \Delta t + o(\Delta t).$$

Moreover, let

$$\lambda_1 = -\lambda,$$

so that

$$\sum_k \lambda_k = 0. \tag{4}$$

Then the Kolmogorov equations (8.15), p. 105 for the transition probabilities $p_n(t) = p_{1n}(t)$ become

$$\frac{d}{dt} p_n(t) = \sum_k \lambda_k p_{kn}(t), \qquad n = 0, 1, 2, \ldots$$

Next we deduce a corresponding differential equation for the generating function $F(t, z)$. Clearly

$$\frac{d}{dt} F(t, z) = \frac{d}{dt} \sum_{n=0}^{\infty} p_n(t) z^n = \sum_{n=0}^{\infty} z^n \frac{d}{dt} p_n(t) = \sum_k \lambda_k \sum_{n=0}^{\infty} p_{kn}(t) z^n \tag{5}$$

(justify the term-by-term differentiation), where $F_k(t, z)$ is the generating function of the random variable $\xi(t)$ for the case of k original particles.[4] But, according to (1), $\xi(t)$ is the sum of k independent random variables, each with generating function $F(t, z)$. Therefore, by formula (6.7), p. 71,

$$F_k(t, z) = [F(t, z)]^k, \qquad k = 0, 1, 2, \ldots \tag{6}$$

(the formula is trivial for $k = 0$). Substituting (6) into (5), we get

$$\frac{d}{dt} F(t, z) = \sum_k \lambda_k [F(t, z)]^k. \tag{7}$$

---

[3] Note that $F_1(t, z) = F(t, z)$, since clearly $p_{1n}(t) = p_n(t)$.

[4] Clearly $F_0(z) \equiv 1$, since new particles cannot be created in the absence of any original particles.

In what follows, we will assume that a given branching process $\xi(t)$ is specified by giving the transition densities λ_k, $k = 0, 1, 2, \ldots$ Let $f(x)$ be the function defined by the power series

$$f(x) = \sum_{k=0}^{\infty} \lambda_k x^k, \tag{8}$$

so that in particular $f(x)$ is analytic for $0 < x < 1$. Then, according to (7), the generating function $F(t, z)$ satisfies a differential equation of the form

$$\frac{dx}{dt} = f(x). \tag{9}$$

Moreover, since $F(0, z) = z$, the generating function $F(t, z)$ coincides for every z in the interval $0 \leqslant z \leqslant 1$ with the solution $x = x(t)$ of (9) satisfying the initial condition

$$x(0) = z. \tag{10}$$

Instead of (9), it is often convenient to consider the equivalent differential equation

$$\frac{dt}{dx} = \frac{1}{f(x)} \tag{11}$$

for the inverse $t = t(x)$ of the function $x = x(t)$. The function satisfying (11) and the initial condition (10) is just

$$t = \int_z^x \frac{du}{f(u)}, \qquad 0 \leqslant x \leqslant 1.$$

***Example* 1.** If

$$\lambda_0 = \lambda, \qquad \lambda_1 = -\lambda,$$
$$\lambda_k = 0, \qquad k = 2, 3, \ldots,$$

then

$$f(x) = \lambda(1 - x)$$

and

$$t = \int_z^x \frac{du}{f(u)} = -\frac{1}{\lambda}[\ln(1 - x) - \ln(1 - z)].$$

Hence $F = F(t, z)$ is such that

$$\ln(1 - F) = -\lambda t + \ln(1 - z),$$

i.e.,

$$F(t, z) = 1 - e^{-\lambda t}(1 - z).$$

The probabilities $p_n(t)$ are found from the expansion

$$F(t, z) = \sum_{n=0}^{\infty} p_n(t) z^n,$$

which in this case implies

$$p_0(t) = 1 - e^{-\lambda t}, \qquad p_1 = e^{-\lambda t},$$
$$p_n(t) = 0, \qquad n = 2, 3, \ldots$$

***Example* 2.** If

$$\lambda_0 = 0, \quad \lambda_1 = -1,$$
$$\lambda_k = \frac{1}{(k-1)k}, \qquad k = 2, 3, \ldots,$$

then

$$f(x) = \sum_{k=0}^{\infty} \lambda_k x^k = \sum_{k=2}^{\infty} \frac{x^k}{k-1} - \sum_{k=1}^{\infty} \frac{x^k}{k}$$
$$= -x \ln(1-x) + \ln(1-x) = (1-x)\ln(1-x),$$

and hence

$$t = \int_z^x \frac{du}{f(u)} = \int_z^x \frac{du}{(1-u)\ln(1-u)} = -\int_{\ln(1-z)}^{\ln(1-x)} \frac{du}{u}$$
$$= -\ln\ln(1-x) + \ln\ln(1-z).$$

It follows that $F = F(t, z)$ is such that

$$\frac{\ln(1-F)}{\ln(1-z)} = e^{-t},$$

i.e.,

$$F(t, z) = 1 - (1-z)^{e^{-t}}.$$

To find the corresponding probabilities $p_n(t)$, we use repeated differentiation:

$$p_0(t) = 0, \qquad p_1(t) = e^{-t}$$
$$p_n(t) = \frac{1}{n!}\frac{d^n F(t, 0)}{dz^n} = \frac{1}{n!}(-1)^{n-1}e^{-t}(e^{-t}-1)\cdots(e^{-t}-n+1),$$
$$n = 2, 3, \ldots$$

Turning to the analysis of the differential equation (9), where $f(x)$ is given by (8), we note that

$$f''(x) = \sum_{k=2}^{\infty} k(k-1)\lambda_k x^{k-2} \geqslant 0 \qquad \text{if} \quad 0 \leqslant x \leqslant 1.$$

Therefore $f(x)$ is concave upward in the interval $0 \leqslant x \leqslant 1$, with a monotonically increasing derivative. Because of (4), $x = 1$ is a root of the equation $f(x) = 0$. This equation can have at most one other root $x = \alpha$ $(0 < \alpha < 1)$. Thus $f(x)$ must behave in one of the two ways shown in Figure 11.

We now study the more complicated case, where $f(x) = 0$ has two roots $x = \alpha$ $(0 < \alpha < 1)$ and $x = 1$, corresponding to two singular integral

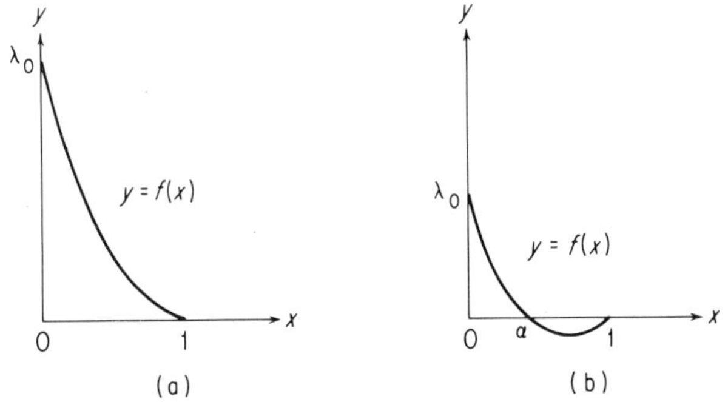

FIGURE 11

curves $x(t) \equiv \alpha$ and $x(t) \equiv 1$ of the differential equations (9) and (11). Consider the integral curve

$$t = \int_z^x \frac{du}{f(u)} \tag{12}$$

going through the point $t = 0$, $x = z$ $(0 \leqslant z < \alpha)$. Since the derivative $f'(\alpha)$ is finite and $f(x) \sim f'(x)(x - \alpha)$ for $x \sim \alpha$, the value of t along the integral curve (12) increases without limit as $x \to \alpha$, but the curve itself never intersects the other integral curve $x(t) \equiv \alpha$. The function $f(x)$ is positive in the interval $0 \leqslant x \leqslant \alpha$, and hence the integral curve $x = x(t)$ increases monotonically as $t \to \infty$, remaining bounded by the value $x = \alpha$. Being a bounded monotonic function, $x(t)$ has a limit

$$\beta = \lim_{t\to\infty} x(t), \qquad z \leqslant \beta \leqslant \alpha.$$

But $f(x)$ approaches a limit $f(\beta)$ as $x \to \beta$, i.e.,

$$f(\beta) = \lim_{t\to\infty} f[x(t)] = \lim_{t\to\infty} x'(t),$$

where $f(\beta)$ must vanish, since otherwise the function

$$x(t) = z + \int_0^t f[x(s)]\, ds$$

would increase without limit as $t \to \infty$. It follows that β is a root of the equation $f(x) = 0$, and hence must coincide with α. Therefore all the integral curves $x = x(t)$ going through the point $x = z$, $0 \leqslant z < \alpha$ for $t = 0$ increase monotonically as $t \to \infty$ and satisfy the condition

$$\lim_{t\to\infty} x(t) = \alpha. \tag{13}$$

The behavior of the integral curves going through the point $x = z$, $\alpha < z < 1$ for $t = 0$ is entirely analogous. The only difference is that $x(t)$ now decreases monotonically, since the derivative $x'(t) = f[x(t)]$ is negative and $f(x) \leqslant 0$ for $\alpha < x < 1$. The behavior of typical integral curves in the interval $0 < z < 1$ is shown in Figure 12, where $0 < z_1 < \alpha < z_2 < 1$.

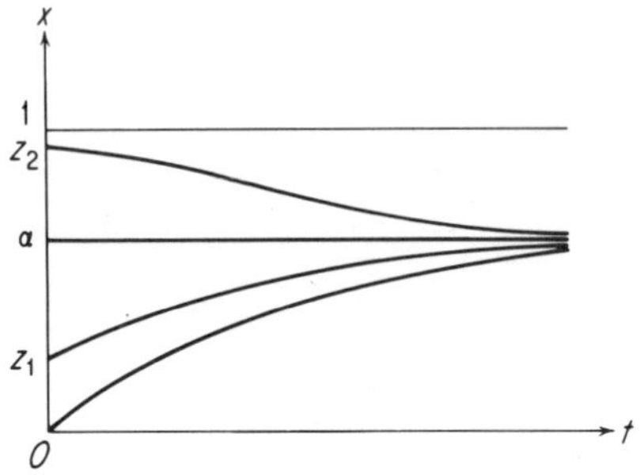

FIGURE 12

The behavior of the integral curves at $z = 1$ warrants special discussion. First we note that in any case $x(t) \equiv 1$ is an integral curve corresponding to $z = 1$. Suppose

$$\int_{x_0}^{1} \frac{dx}{f(x)} = -\infty \tag{14}$$

for some x_0, $\alpha < x_0 < 1$.[5] Then an arbitrary integral curve of the form

$$t = t_0 + \int_{x_0}^{x} \frac{du}{f(u)}, \qquad 0 \leqslant x < 1, \tag{15}$$

going through some point (t_0, x_0), decreases without limit as $x \to 1$, i.e.,

$$t = t_0 + \int_{x_0}^{x} \frac{du}{f(u)} \to -\infty$$

as $x \to 1$. This shows that given any $t_0 > 0$, the equation

$$t(z) = t_0 + \int_{x_0}^{z} \frac{du}{f(u)} = 0$$

holds for some $x = z$, $\alpha < z < 1$. Hence every integral curve intersects the axis $t = 0$ in a point $(0, z)$ such that $\alpha < z < 1$ (see Figure 13). It follows that in this case $x(t) \equiv 1$ is the unique integral curve going through the point $(0, 1)$.

On the other hand, suppose

$$\int_{x_0}^{1} \frac{dx}{f(x)} > -\infty. \tag{16}$$

Then for sufficiently large t_0, the integral curve (15) intersects the integral curve $x(t) \equiv 1$, and is in fact tangent to it at the point $(\tau, 1)$ where

$$\tau = t_0 + \int_{x}^{1} \frac{dx}{f(x)}$$

[5] This is always the case if $f'(1) < \infty$ (why?).

(see Figure 13). In this case, there is a whole family of integral curves $x_\tau(t)$ going through the point $(0, 1)$, where each $x_\tau(t)$ is parameterized by the appropriate value of $\tau \geqslant 0$. Among these integral curves, the curve $x_0(t)$ shown in the figure, corresponding to the value $\tau = 0$, has the property of lying below all the other integral curves, i.e.,

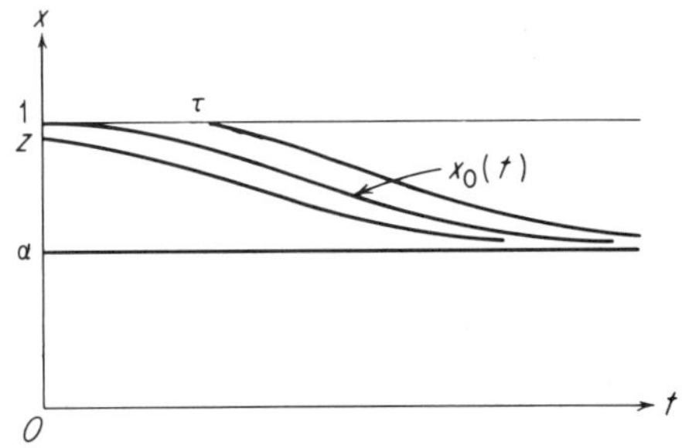

FIGURE 13

$$x_0(t) \leqslant x_\tau(t), \qquad 0 \leqslant t < \infty.$$

This is explained by the fact that the solution of our differential equation is unique in the region $0 \leqslant x < 1$, $0 \leqslant t < \infty$, so that the integral curves do not intersect in this region. It is also easy to see that the integral curve $x_0(t)$ is the limit of the integral curves $x(t, z)$ lying below it and passing through points $(0, z)$ such that $0 \leqslant z < 1$. In other words,[6]

$$x_0(t) = \lim_{z \to 1} x(t, z). \tag{17}$$

The above analysis of the differential equation (9) has some interesting implications for the corresponding branching process $\xi(t)$. In general, there is a positive probability that no particles at all are present at a given time t. Naturally, this cannot happen if $\lambda_0 = 0$, since then particles can only be "created" but not "annihilated." Clearly, the probability of all particles having disappeared after time t is

$$p_0(t) = F(t, 0)$$

if there is only one particle originally present at time $t = 0$, and

$$p_{k0}(t) = [F(t, 0)]^k = [p_0(t)]^k$$

if there are k particles at time $t = 0$. The function $p_0(t)$ is the solution of the differential equation (9) corresponding to the parameter $z = 0$:

$$\frac{dp_0(t)}{dt} = f[p_0(t)], \qquad p_0(0) = 0.$$

As already shown, this solution asymptotically approaches some value $p_0 = \alpha$ as $t \to \infty$, where α is the smaller root of the equation $f(x) = 0$ [recall (13)]. Thus $p_0 = \alpha$ is the *extinction probability* of the branching process $\xi(t)$, i.e., the probability that all the particles will eventually disappear. If the function $f(x)$ is positive in the whole interval $0 \leqslant x < 1$, the extinction probability equals 1.

---

[6] Note that $x(t, z) \equiv F(t, z)$ for $t \geqslant 0$, $0 \leqslant z < 1$.

There is also the possibility of an "explosion" in which infinitely many particles are created. The probability of an explosion occurring by time t is just

$$p_\infty(t) = 1 - \mathbf{P}\{\xi(t) < \infty\} = 1 - \sum_{n=0}^{\infty} \mathbf{P}\{\xi(t) = n\}$$

$$= 1 - \sum_{n=0}^{\infty} p_n(t) = 1 - \lim_{z\to 1} F(t, z).$$

In the case where $x(t) \equiv 1$ is the unique integral curve of (9) passing through the point (0, 1), we clearly have

$$\lim_{z\to 1} F(t, z) = 1.$$

Therefore $p_\infty(t) = 0$ for arbitrary t if (14) holds, and the probability of an explosion ever occurring is 0. However, if (16) holds, we have (17) where $x_0(t)$ is the limiting integral curve described above and shown in Figure 13. In this case,

$$p_\infty(t) = 1 - x_0(t) > 0$$

and there is a positive probability of an explosion occurring.

PROBLEMS

1. A cosmic ray shower is initiated by a single particle entering the earth's atmosphere. Find the probability $p_n(t)$ of n particles being present after time t if the probability of each particle producing a new particle in a small time interval Δt is $\lambda\Delta t + o(\Delta t)$.

Hint. $\lambda_1 = -\lambda$, $\lambda_2 = \lambda$.

Ans. $p_n(t) = e^{-\lambda t}(1 - e^{-\lambda t})^{n-1}$, $n \geqslant 1$.

2. Solve Problem 1 if each particle has probability $\lambda\Delta t + o(\Delta t)$ of producing a new particle and probability $\mu\Delta t + o(\Delta t)$ of being annihilated in a small time interval Δt.

Hint. $\lambda_0 = \mu$, $\lambda_1 = -(\lambda + \mu)$, $\lambda_2 = \lambda$.

Ans. $p_0(t) = \mu\gamma$, $p_n(t) = (1 - \lambda\gamma)(1 - \mu\gamma)(\lambda\gamma)^{n-1}$ $(n \geqslant 1)$,

where

$$\gamma = \begin{cases} \dfrac{1 - e^{(\lambda-\mu)t}}{\mu - \lambda e^{(\lambda-\mu)t}} & \text{if } \lambda \neq \mu, \\[2ex] \dfrac{t}{1 + \lambda t} & \text{if } \lambda = \mu. \end{cases}$$

3. Find the extinction probability p_0 of the branching process in the preceding problem.

Ans. $$p_0 = \begin{cases} \dfrac{\mu}{\lambda} & \text{if} \quad \mu < \lambda, \\ 1 & \text{if} \quad \mu \geqslant \lambda. \end{cases}$$

Appendix 4

PROBLEMS OF OPTIMAL CONTROL

As in Sec. 15, consider a physical system which randomly changes its state at the times $t = 1, 2, \ldots$, starting from some initial state at time $t = 0$. Let $\varepsilon_1, \varepsilon_2, \ldots$ be the possible states of the system, and $\xi(t)$ the state of the system at time t, so that the evolution of the system in time is described by the consecutive transitions

$$\xi(0) \to \xi(1) \to \xi(2) \cdots .$$

We will assume that $\xi(t)$ is a Markov chain, whose transition probabilities p_{ij}, $i, j = 1, 2, \ldots$ depend on a "control parameter" chosen step by step by an external "operator." More exactly, if the system is in state ε_i at any time n and if d is the value of the control parameter chosen by the operator, then

$$p_{ij} = p_{ij}(d)$$

is the probability of the system going into the state ε_j at the next step. The set of all possible values of the control parameter d will be denoted by D.

We now pose the problem of controlling this "guided random process" by bringing the system into a definite state, or more generally into one of a given set of states E, after a given number of steps n. Since the evolution of the process $\xi(t)$ depends not only on the control exerted by the operator, but also on chance, there is usually only a definite probability P of bringing the system into one of the states of the set E, where P depends on the "control program" adopted by the operator. We will assume that every such control program consists in specifying in advance, for all ε_i and $t = 0, \ldots, n - 1$, the parameter

$$d = d(\varepsilon_i, t)$$

to be chosen if the system is in the state ε_i at the time t. In other words, the whole control program is described by a *decision rule*, i.e., a function of two variables

$$d = d(x, t),$$

where x ranges over the states $\varepsilon_1, \varepsilon_2, \ldots$ and t over the times $0, \ldots, n-1$. Thus the probability of the system going into the state ε_j at time $k+1$, given that it is in the state ε_i at time k, is given by

$$p_{ij} = p_{ij}(d), \qquad d = d(\varepsilon_i, k).$$

By the same token, the probability of the system being guided into one of the states in E depends on the choice of the control program, i.e., on the decision rule $d = d(x, t)$, so that

$$P = P(d).$$

Control with a decision rule $d^0 = d^0(x, t)$ will be called *optimal* if

$$P(d^0) = \max_d P(d),$$

where the maximum is taken with respect to all possible control programs, i.e., all possible decision rules $d = d(x, t)$. Our problem will be to find this optimal decision rule d^0, thereby maximizing the probability

$$P(d) = \mathbf{P}\{\xi(n) \in E\}$$

of the system ending up in one of the states of E after n steps.

We now describe a multistage procedure for finding d^0. Let

$$P(k, i, d) = \mathbf{P}\{\xi(n) \in E \mid \xi(k) = \varepsilon_i\}$$

be the probability that after occupying the state ε_i at the kth step, the system will end up in one of the states of the set E after the remaining $n-k$ steps (it is assumed that some original choice of the decision rule $d = d(x, t)$ has been made). Then clearly

$$P(k, i, d) = \sum_j p_{ij}(d) P(k+1, j, d). \tag{1}$$

This is a simple consequence of the total probability formula, since at the $(k+1)$st step the system goes into the state ε_j with probability $p_{ij}(d)$, $d = d(\varepsilon_i, k)$, whence with probability $P(k+1, j, d)$ it moves on ($n-k-1$ steps later) to one of the states in the set E.

For $k = n-1$, formula (1) involves the probability

$$P(n, j, d) = \begin{cases} 1 & \text{if } \varepsilon_j \in E, \\ 0 & \text{otherwise,} \end{cases} \tag{2}$$

and hence

$$P(n-1, i, d) = \sum_{j:\, \varepsilon_j \in E} p_{ij}(d), \tag{3}$$

where the summation is over all j such that the state ε_j belongs to the given set E. Obviously, $P(n-1, i, d)$ does not depend on values of the control parameter other than the values $d(\varepsilon_i, n-1)$ chosen at the time $n-1$. Letting d^0 denote the value of the control parameter at which the function (3) takes its maximum,[1] we have

$$P^0(n-1, i) = P(n-1, i, d^0) = \max_{d \in D} P(n-1, i, d). \tag{4}$$

Clearly, there is a value $d^0 = d^0(\varepsilon_i, n-1)$ corresponding to every pair $(\varepsilon_i, n-1)$, $i = 1, 2, \ldots$

For $k = n-2$, formula (1) becomes

$$P(n-2, i, d) = \sum_j p_{ij}(d)P(n-1, j, d).$$

Here the probabilities $p_{ij}(d)$ depend only on the values $d = d(\varepsilon_i, n-2)$ of the decision rule $d = d(x, t)$ chosen at time $n-2$, while the probabilities $P(n-1, j, d)$ depend only on the values $d = d(\varepsilon_j, n-1)$ chosen at time $n-1$. Suppose we "correct" the decision rule $d = d(x, t)$ by replacing the original values $d(\varepsilon_j, n-1)$ by the values $d^0(\varepsilon_j, n-1)$ just found. Then the corresponding probabilities $P(n-1, j, d)$ increase to their maximum values $P^0(n-1, j)$, thereby increasing the probability $P(n-2, i, d)$ to the value

$$P(n-2, i, d) = \sum_j p_{ij}(d)P^0(n-1, j). \tag{5}$$

Clearly, (5) depends on the decision rule $d = d(t, x)$ only through the dependence of the transition probabilities $p_{ij}(d)$ on the values $d = d(\varepsilon_i, n-2)$ of the control parameter at time $n-2$. Again letting d^0 denote the value of the control parameter at which the function (5) takes its maximum, we have

$$P^0(n-2, i) = P(n-2, i, d^0) = \max_{d \in D} P(n-2, i, d).$$

As before, there is a value $d^0 = d^0(\varepsilon_i, n-2)$ corresponding to every pair $(\varepsilon_i, n-2)$, $i = 1, 2, \ldots$ Suppose we "correct" the decision rule $d(x, t)$ by setting

$$d(x, t) = d^0(x, t) \tag{6}$$

for $t = n-2, n-1$ and all $x = \varepsilon_1, \varepsilon_2, \ldots$ Then clearly the probabilities $P(k, i, d)$ take their maximum values $P^0(k, i)$ for $i = 1, 2, \ldots$ and $k = n-2$, $n-1$. Correspondingly, formula (1) becomes

$$P(n-3, i, d) = \sum_j p_{ij}(d)P(n-2, j, d) = \sum_j p_{ij}(d)P^0(n-2, j),$$

and this function of the control parameter d takes its maximum for some $d^0 = d^0(\varepsilon_i, n-3)$. We can then, once again, "correct" the decision rule

[1] It will be assumed that this maximum and the others considered below exist.

$d = d(x, t)$ by requiring (6) to hold for $t = n - 3$ and all $x = \varepsilon_1, \varepsilon_2, \ldots$, as well as for $t = n - 2, n - 1$ and all $x = \varepsilon_1, \varepsilon_2, \ldots$

Continuing this step-by-step procedure, after $n - 1$ steps we eventually get the optimal decision rule $d = d^0(x, t)$, defined for $t = 0, \ldots, n - 1$ and all $x = \varepsilon_1, \varepsilon_2, \ldots$, such that the probability $P(d) = P(0, i, d)$ satisfying the initial condition $\xi(0) = \varepsilon_i$ achieves its maximum value. At the $(n - k)$th step of this procedure of "successive corrections," we find the value $d^0 = d^0(\varepsilon_i, k)$ maximizing the function

$$P(k, i, d) = \sum_j p_{ij}(d) P^0(k + 1, j),$$

where $P^0(k + 1, j)$ is the maximum value of the probability $P(k + 1, j, d)$. Carrying out this maximization, we get *Bellman's equation*[2]

$$P^0(k, i) = \max_{d \in D} \sum_j p_{ij}(d) P^0(k + 1, j),$$

which summarizes the whole procedure just described.

***Example* 1.** Suppose there are just two states ε_1 and ε_2, and suppose the transition probabilities are continuous functions of the control parameter in the intervals

$$\alpha_1 \leqslant p_{11}(d) \leqslant \beta_1, \qquad \alpha_2 \leqslant p_{21}(d) \leqslant \beta_2.$$

What is the optimal decision rule maximizing the probability of the system, initially in the state ε_1, going into the state ε_1 two steps later?

Solution. In this case,

$$P^0(1, 1) = \beta_1, \qquad P^0(1, 2) = \beta_2,$$

$$P^0(0, 1) = \max_d \, [p_{11}(d)\beta_1 + p_{12}(d)\beta_2] = \max_d \, [p_{11}(d)(\beta_1 - \beta_2) + \beta_2].$$

If the system is initially in the state ε_1, then clearly we should maximize the transition probability p_{11} (by choosing $p_{11} = \beta_1$) if $\beta_1 > \beta_2$, while maximizing the transition probability $p_{12} = 1 - p_{11}$ (by choosing $p_{11} = \alpha_1$) if $\beta_1 < \beta_2$.[3] There is an analogous optimal decision rule for the case where the initial state of the system is ε_2.

***Example* 2 (*The optimal choice problem*).** Once again we consider the optimal choice problem studied on pp. 28–29 and 86–87, corresponding to

---

[2] In keeping with (2)–(4), we have

$$P^0(n, j) = \begin{cases} 1 & \text{if } \varepsilon_j \in E, \\ 0 & \text{otherwise.} \end{cases}$$

[3] Clearly, any choice of p_{11} in the interval $\alpha_1 \leqslant p_{11} \leqslant \beta_1$ is optimal if $\beta_1 = \beta_2$.

a Markov process $\xi(t)$ with transition probabilities

$$p_{ij} = \begin{cases} 0 & \text{if } i \geqslant j, \\ \dfrac{i}{(j-1)j} & \text{if } i < j \leqslant m, \\ \dfrac{i}{m} & \text{if } j = m+1, \end{cases} \tag{7}$$

where, as on p. 28, choice of an object better than all those previously inspected causes the process $\xi(0) \to \xi(1) \to \xi(2) \to \cdots$ to terminate. In each of the states $\varepsilon_1, \ldots, \varepsilon_m$ (whose meaning is explained on p. 86), the observer decides whether to terminate or to continue the process of inspection. The decision to terminate, if taken in the state ε_i, is described formally by the transition probabilities

$$p_{ij} = \begin{cases} 1 & \text{if } i = j, \\ 0 & \text{if } i \neq j, \end{cases} \tag{8}$$

while the decision to continue corresponds to the transition probabilities (7). Hence we are dealing with a "guided Markov process," whose transition probabilities p_{ij} depend on the observer's decision. Here the control parameter d takes only two values, 0 and 1 say, where 0 corresponds to stopping the process and 1 to continuing it. Thus (8) gives the probabilities $p_{ij}(0)$ and (7) the probabilities $p_{ij}(1)$.

Every inspection plan is described by a decision rule $d = d(x)$, $x = \varepsilon_1, \ldots, \varepsilon_m$, which specifies in advance for each of the states $\varepsilon_1, \ldots, \varepsilon_m$ whether inspection should be continued or terminated by selecting the last inspected object. The problem consists of finding an inspection plan, or equivalently a decision rule $d = d(x)$, $x = \varepsilon_1, \ldots, \varepsilon_m$, maximizing the probability of selecting the very best of all m objects. This probability is just

$$P = \sum_i \frac{i}{m} p_i, \tag{9}$$

where i/m is the probability that the ith inspected object is the best (recall p. 29), p_i is the probability that the process will stop in the state ε_i, and the summation is over all the states ε_i in which the decision rule $d = d(x)$ calls for the process to stop.

To find the optimal decision rule $d^0 = d^0(x)$ maximizing (9), we consider the probability $P(k, d)$ of selecting the best object, given that the number of previously inspected objects is no less than k, i.e., given that the process $\xi(t)$ actually occupies the state ε_k. By the total probability formula, we have

$$P(k, d) = \sum_{j=k}^{m} p_{kj}(d) P(j, d). \tag{10}$$

Clearly, if the process occupies the state ε_m, then the mth object is the best of the first m objects inspected and hence automatically the best of all objects. Therefore the optimal value of the decision rule $d = d(x)$ for $x = \varepsilon_m$ is just $d^0(\varepsilon_m) = 0$, and $P(m, d) = 1$ for this value. It follows from (9) and (10) that

$$P(m-1, d) = \begin{cases} \dfrac{m-1}{m} & \text{if } d(\varepsilon_{m-1}) = 0, \\[2ex] \dfrac{m-1}{(m-1)m} & \text{if } d(\varepsilon_{m-1}) = 1 \end{cases} \tag{11}$$

is the probability of choosing the best object, given that the process stops in the state ε_m and the number of previously inspected objects is no less than $m - 1$. Moreover, (11) implies that the optimal value of the decision rule $d = d(x)$ for $x = \varepsilon_{m-1}$ is $d^0(\varepsilon_{m-1}) = 0$, and that

$$P^0(m-1) = \frac{m-1}{m}$$

Now suppose the optimum values of the decision rule $d = d(x)$ are all zero for $x = \varepsilon_k, \ldots, \varepsilon_m$, corresponding to the fact that the process is terminated in any of the states $\varepsilon_k, \ldots, \varepsilon_n$. Then what is the optimal value $d^0(\varepsilon_{k-1})$? To answer this question, we note that (9) and (10) imply that

$$P(k-1, d) = \begin{cases} \dfrac{k-1}{m} & \text{if } d(\varepsilon_{k-1}) = 0, \\[2ex] \dfrac{k-1}{(k-1)k}\dfrac{k}{m} + \dfrac{k-1}{k(k+1)}\dfrac{k+1}{m} + \cdots + \dfrac{k-1}{(m-1)m} \cdot 1 & \text{if } d(\varepsilon_{k-1}) = 1 \end{cases}$$

is the probability of choosing the best object, given that the process stops in the states $\varepsilon_k, \ldots, \varepsilon_m$ and the number of previously inspected objects is no less than $k - 1$. It follows that the optimal value of the decision rule $d = d(x)$ for $x = \varepsilon_{k-1}$ is

$$d^0(\varepsilon_{k-1}) = \begin{cases} 0 & \text{if } \dfrac{1}{k-1} + \dfrac{1}{k} + \cdots + \dfrac{1}{m-1} \leqslant 1, \\[2ex] 1 & \text{otherwise.} \end{cases} \tag{12}$$

Moreover, it is easy to see that the optimal decision rule $d^0 = d^0(x)$ has the structure

$$d^0(x) = \begin{cases} 0 & \text{if } x = \varepsilon_{m_0}, \ldots, \varepsilon_m, \\ 1 & \text{if } x = \varepsilon_1, \ldots, \varepsilon_{m_0-1}, \end{cases}$$

where m_0 is some integer. Thus the optimal selection procedure consists in continuing inspection until the appearance of an object numbered $k \geqslant m_0$ which is better than all previously inspected objects. According to (12), m_0 is the largest positive integer such that

$$\frac{1}{m_0} + \frac{1}{m_0 + 1} + \cdots + \frac{1}{m - 1} > 1. \tag{13}$$

PROBLEMS

1. In Example 2, prove that

$$m_0 \approx \frac{m}{e} \tag{14}$$

if m is large, where $e = 2.718\ldots$ is the base of the natural logarithms.

Hint. Use an integral to estimate the left-hand side of (13).

2. Find the exact value of m_0 for $m = 50$. Compare the result with (14).

3. Consider a Markov chain with two states ε_1 and ε_2 and transition probabilities $p_{ij}(d)$ depending on a control parameter d taking only two values 0 and 1. Suppose

$$p_{11}(0) = \tfrac{1}{5}, \quad p_{21}(0) = \tfrac{4}{5}, \quad p_{11}(1) = \tfrac{2}{5}, \quad p_{21}(1) = \tfrac{3}{5}.$$

What is the optimal decision rule maximizing the probability of the system initially in the state ε_1, going into the state ε_2 three steps later? What is this maximum probability?

BIBLIOGRAPHY

Chung, K. L., *Markov Chains with Stationary Transition Probabilities*, Springer-Verlag, Berlin (1960).

Chung, K. L., *A Course in Probability Theory*, Harcourt, Brace and World, Inc., New York (1968).

Fano, R. M., *Transmission of Information*, John Wiley and Sons, Inc., New York (1961).

Feller, W., *An Introduction to Probability Theory and Its Applications*, John Wiley and Sons, Inc., New York, *Volume I*, third edition (1968), *Volume II* (1966).

Gnedenko, B. V., *The Theory of Probability*, fourth edition (translated by B. D. Seckler), Chelsea Publishing Co., New York (1967).

Harris, T. E., *The Theory of Branching Processes*, Prentice-Hall, Inc., Englewood Cliffs, N.J. (1963).

Loève, M., *Probability Theory*, third edition, D. Van Nostrand Co., Inc., Princeton, N.J. (1963).

McKinsey, J. C. C., *Introduction to the Theory of Games*, McGraw-Hill Book Co., Inc., New York (1952).

Parzen, E., *Modern Probability Theory and Its Applications*, John Wiley and Sons, Inc., New York (1960).

Uspensky, J. V., *Introduction to Mathematical Probability*, McGraw-Hill Book Co., Inc., New York (1937).

Williams, J. D., *The Compleat Strategyst*, McGraw-Hill Book Co., Inc., New York (1954).

INDEX

L

M

N

O

P

Q

R